The Gentle Art of Mathematics

The Gentle Art of Mathematics

DAN PEDOE

With Drawings by Griselda El Tayeb

DOVER PUBLICATIONS, INC.
NEW YORK

Published in Canada by General Publishing Company, Ltd., 30 Lesmill Road, Don Mills, Toronto, Ontario.

This Dover edition, first published in 1973, is an unabridged, slightly corrected republication of the work originally published in 1959 by The Macmillan Company.

International Standard Book Number: 0-486-22949-1
Library of Congress Catalog Card Number: 73-77445

Manufactured in the United States of America
Dover Publications, Inc.
180 Varick Street
New York, N. Y. 10014

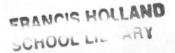

PREFACE

This book is intended for the many people who would like to know what mathematics is about, especially modern mathematics. Mathematics has many important practical applications, but should not be of interest only to the scientist. There is much in mathematical thought which should interest the arts student, and much which is beautiful and should interest everyone.

Those who profess mathematics do so because they enjoy it. To understand and share in this enjoyment, the reader is invited to follow some of the arguments given in the following pages. Explanations have been made as detailed as is reasonably possible. Although tricks and puzzles which do not involve some general principle have, on the whole, been omitted, much that is recreational will be found here. For a complete understanding of this book the reader should, of course, verify the calculations and ponder the arguments as he comes across them; but even if he goes through the book like the gentleman "who read Euclid, all except the A's and the B's and the pictures of scratches and scrawls", he will find enough of a purely descriptive nature to provide entertainment and instruction.

The author has selected those topics for discussion which he feels are of interest, but are not easily accessible elsewhere. This will explain the omission of the differential and integral calculus, and the inclusion of chapters on probability (chance and choice) and symbolic logic (automatic thinking). The expert reader will feel, no doubt, that many of the chapters could have easily been extended; but he must also be aware that there is a close relation between the size of a book and its cost.

Perhaps a novel feature of this book is the non-inclusion of any quotations from *Alice in Wonderland* or *Through the Looking-Glass*, although some characteristic examples from *Symbolic Logic* by the Reverend Charles Lutwidge Dodgson have been included in the chapter called "Automatic Thinking". The Lewis Carroll attitude to mathematics, delightful though it is, has probably had an inhibiting effect on many potential mathematicians. It has made them feel that mathematics is a subject which only clever, witty people can learn and understand.

It is true that some mathematicians are both witty and clever; but not all mathematicians are witty and clever. The only apparatus a mathematician needs is a capacity for logical thought, and most people have this. Augustus De Morgan, who was a contemporary of Lewis Carroll's, said: "I divide the illogical—I mean people who have not that amount of natural use of sound inference which is really not uncommon—into three classes:—First class, three varieties: the Niddy, the Noddy and the Noodle. Second class, three varieties: the Niddy-Noddy, the Niddy-Noodle, and the Noddy-Noodle. Third class, undivided: the Niddy-Noddy-Noodle."

This book is for the many who are neither Niddys, Noddys, Noodles nor any combination thereof.

I am very grateful to Dr. Doris Lee and Sir Graham Sutton, F.R.S. for their comments on the *ms.* of this book. The Warburg Institute was very helpful in providing two illustrations, and the Pepys Librarian of Magdalene College, Cambridge provided the text of the two Newton letters, thus putting me under an even greater debt to my old college. My wife, besides encouraging me constantly, resuscitated the ancient rhyme for π, and produced one of the most beautiful and difficult drawings. Finally, the English Universities Press has met my wishes most admirably in their production of this volume.

D.P.

Khartoum, October, 1957

CONTENTS

7

CHAPTER I

MATHEMATICAL GAMES

MANKIND has always been fascinated by the ordinary integers, the *natural numbers* 1, 2, 3, 4, 5, 6, 7, 8, 9, ... At a very early age the normally endowed human being is made aware that he possesses 5 fingers on each hand and 5 toes on each foot. Sooner or later he believes that he himself is a unique individual; that he represents the number 1. The biological significance of the number 2 becomes only too clear. A mystic significance grows around the number 3. To bridge players 4 is a basic integer, and similarly 5, 6, 7, 8 and 9 all have their devotees. A favourite pastime in the not too distant past was to attach numbers to the letters of the alphabet so that, when you added up the numbers corresponding to the letters, the name of your enemy was shown to add up to the Number of the Beast, which is 6 6 6, according to Revelations. Much has been written on the association of integers with remarkable events. Good use was made of this knowledge by a Mr. Galloway, a Fellow of the Royal Society, when the Council of that august body first resolved to restrict the number of yearly admissions to the Society to fifteen men of science and noblemen *ad libitum*. Why fifteen, asked Mr. Galloway? He then continued, with true Victorian thoroughness:

Was it because fifteen is seven and eight, typifying the Old Testament Sabbath, and the New Testament day of the resurrection following? Was it because Paul strove fifteen days against Peter, proving that he was a doctor both of the Old and the New Testament? Was it because the prophet Hosea bought a lady for fifteen pieces of silver? Was it because, according to Micah, seven shepherds and eight chiefs should waste the Assyrians? Was it because Ecclesiastes commands equal reverence to be given to both Testaments—such was the interpretation—in the words "Give a portion to seven, and also to eight"? Was it because the waters of the deluge rose fifteen cubits above the mountains?—or because they lasted fifteen decades of days? Was it because Ezekiel's temple had fifteen steps? Was it because Jacob's ladder has been supposed to have had fifteen steps? Was it because fifteen years were added to the life of Hezekiah? Was it because the feast of unleavened bread was on the fifteenth day of the month? Was it because the scene of the Ascension was fifteen stadia from Jerusalem? Was it because the stone-masons and porters employed on Solomon's temple amounted to fifteen myriads?

As this is not a book devoted only to mathematical curiosities we shall not spend too much time on these *shallow numerists*, which is what Cocker, the author of a famous 17th-century arithmetic book, called these numbermongers. In this chapter we consider the way in which ordinary integers can be represented, and describe a number of mathematical problems and games which arise from this representation.

The numbers we use in ordinary life are expressed in the *scale of ten*. Perhaps there is no need to stress this. *Twenty* means *twice* ten, *thirty* means *three* times ten, and so on. We use words like a *hundred* for ten times ten, a *thousand* for ten times a hundred, a *million* for a thousand thousand. The mathematician prefers to use unambiguous symbols instead of words, if he can, and writes

$$100 = (10) . (10) = 10^2,$$

where . is used as the symbol for multiplication, and the index 2 is used to show that 10 is multiplied by itself twice. With this notation

$$1000 = 10 . 10 . 10 = 10^3,$$

and

$$1,000,000 = 10^6.$$

The number 9824, say, stands for

$$9 . 10^3 + 8 . 10^2 + 2 . 10 + 4.$$

The position of the digits as we move from left to right in a number indicates which powers of ten are involved, and when we assess any number with a largish set of digits, like 54623108, we do a rapid mental calculation. We begin at the right, and mark off the digits in sets of three, so that we have

$$54, 623, 108,$$

and then we know that the number is fifty-four million, six hundred and twenty-three thousand, one hundred and eight. Or, since there are 8 digits involved, we also know that the number can be represented, using powers of 10, as

$$5 . 10^7 + 4 . 10^6 + 6 . 10^5 + 2 . 10^4$$
$$+ 3 . 10^3 + 1 . 10^2 + 0 . 10 + 8.$$

We note that the highest index involved is one less than the total number of digits.

How can we represent a general number with n digits? Here n represents any integer, such as 1, 2, 3, ... We require n different symbols for the n digits. There are advantages in using a single symbol, with a digit *suffix* to distinguish one symbol from any other. We could use

$$a_0, a_1, a_2, a_3, \ldots\ldots, a_{n-1},$$

to represent n digits. We have brought in the *cipher*, or zero, 0, for use with the digits. We shall see that it is useful. The number with n digits will be written

$$a_{n-1}\, a_{n-2}\, a_{n-3} \ldots\ldots a_3\, a_2\, a_1\, a_0$$

in the ordinary way, where the *position* of each digit is significant, as in ordinary numbers. We then know that the number represented is

$$10^{n-1} \cdot a_{n-1} + 10^{n-2} \cdot a_{n-2} + \ldots + 10^2 \cdot a_2 + 10^1 \cdot a_1 + a_0.$$

The suffixes have indicated the power of 10 by which the corresponding digit is to be multiplied.

We make use of this discussion to solve a digital problem which runs as follows:

Find the smallest integer which is such that if the digit on the extreme left is transferred to the extreme right, the new number so formed is one and a half times the original number.*

If the number were 5364, the digit on the extreme left is 5, and after transfer to the extreme right the number is 3645. But this is not the solution. The interest of this problem lies in the fact that the answer is so very large, and in the even more pertinent fact that a method of solution which is not systematic and logical will hardly obtain the result in a finite time! All the same, the reader is invited to have a go before he studies the solution. Since answers are usually given in examinations on the higher mathematics, we state the result: the smallest number satisfying the conditions is

$$1, 176, 470, 588, 235, 294.$$

When we transfer the digit from the extreme left to the extreme right we have

$$1, 764, 705, 882, 352, 941,$$

and a moment's calculation will show that this second number is indeed one and a half times the first one.

* Set as a Chrismas teaser by Dr. J. Bronowski in the *New Statesman and Nation*, December 24, 1949.

The solution we now describe is fairly long, but all the steps in it are simple ones. A much shorter solution will follow, and the reader may turn to the shorter one first, if he so desires. We first write down the unknown number, and it is convenient to assume that it is a number of $n + 1$ digits, rather than a number of n digits. The second assumption is no more general than the first. The number will be

$$a_n \, a_{n-1} \, a_{n-2} \, \ldots \ldots \, a_2 \, a_1 \, a_0 \, ,$$

and the digit on the extreme left being a_n, when we transfer it to the extreme right we obtain the number

$$a_{n-1} \, a_{n-2} \, \ldots \ldots \ldots \, a_2 \, a_1 \, a_0 \, a_n \, .$$

Since we do not wish to be restricted by the positions of the various a's in each number, we write the given number in the form

$$10^n \, . \, a_n + 10^{n-1} \, . \, a_{n-1} + \ldots . + 10 \, . \, a_1 + a_0 \, .$$

We note that each of the various a's is less than or equal to 9. This point will be significant later. When we have transferred the digit, the new number is

$$10^n \, . \, a_{n-1} + 10^{n-1} \, . \, a_{n-2} + \ldots + 10^2 \, . \, a_1 + 10 \, . \, a_0 + a_n \, .$$

By the given conditions of the problem, this new number is $3/2$ times the one we started with, and so we arrive at the equation:

$$3(10^n \, . \, a_n + 10^{n-1} \, . \, a_{n-1} + \ldots . + a_0)$$
$$= 2(10^n \, . \, a_{n-1} + 10^{n-1} \, . \, a_{n-2} + \ldots + 10 \, . \, a_0 + a_n).$$

We notice that a_n occurs on both sides of this equation, and all that we do now is to collect together the terms containing a_n on to one side, using the elementary rules of algebra "taking over to the other side changes plus into minus", and so on. No higher mathematics is involved, so far. We obtain

$$(3 \, . \, 10^n - 2) \, a_n$$
$$= (2 \, . \, 10 - 3) \, (10^{n-1} \, . \, a_{n-1} + 10^{n-2} \, . \, a_{n-2} + \ldots + a_0),$$

or

$$(3 \, . \, 10^n - 2) \, a_n = 17(10^{n-1} \, . \, a_{n-1} + 10^{n-2} \, . \, a_{n-2} + \ldots + a_0).$$

This rearrangement has been purely mechanical, guided, of course, by mathematical intuition. But no logical *argument* has entered into the scheme. We now begin! The expressions on each side of the equation are integers. We know that integers can be factorised in only one way into their lowest factors, and these integers, which cannot be factorised any more, are called *prime numbers*. For example 30 = 6 . 5, and 5 cannot be factorised any more, and is prime, but 6 = 3 . 2, and so 30 = 5 . 3 . 2, when factorised into prime numbers. The first primes are 2, 3, 5, 7, 11, 13, 17, ... The prime integer 17 occurs in our equation, on the right-hand side. When the numbers on the left-hand side are factorised, 17 must occur amongst the factors. But a_n, as we remarked earlier, is a digit lying between 1 and 9. The prime factor 17 must therefore arise from the factorisation of the number $3 . 10^n - 2$. In other words,

$$3 . 10^n - 2 \text{ is exactly divisible by 17.}$$

It is this mathematical deduction which enables us to solve our problem.

Since 2 is not divisible by 17, an assertion equivalent to the one above is:

$$3 . 10^n \text{ on division by 17 leaves remainder 2.}$$

It will be remembered that n is one of the things we are trying to find, since the unknown number contains $n + 1$ digits. To find n, all that we have to do is to carry out a long-division sum, until we obtain the remainder 2. We begin with $n = 1$, of course, and see that it will not do, since $3 . 10 = 30$ does not leave remainder 2 on division by 17. When we try $n = 2$, so that we are dividing $3 . 10^2 = 300$ by 17, we merely add another 0 to the number we are dividing. Hence we divide 3 0 0 0 0 0 . . . . . with an indefinite number of 0's until we obtain the remainder 2. The actual long-division is shown below.

We find that the smallest value of n which will do is $n = 15$. That is, we find that $3 . 10^{15}$ is divisible by 17 with remainder 2, or

$$3 . 10^{15} - 2 \text{ is exactly divisible by 17.}$$

The quotient, as the working shows, is 176470588235294. If we look back at the equation which first started us on this division, we see that, since $n = 15$,

$$10^{14} . a_{14} + 10^{13} . a_{13} + \ldots + a_0$$
$$= \frac{1}{17}(3 . 10^{15} - 2) a_{15}$$
$$= 176470588235294 \, a_{15} .$$

Now, the number on the left, if we add $10^{15} . a_{15}$ to it, is the number we are looking for! To make it as small as possible we must take $a_{15} = 1$, and then the number is

$$1, 176, 470, 588, 235, 294.$$

The details of the long division are now given:

```
17 )  3 0 0 0 0 0 0 0 0 0 0 0 0 0 0 0 0   ( 176470588235294
       1 7
       ─────
         1 3 0
         1 1 9
         ─────
           1 1 0
           1 0 2
           ─────
             8 0
             6 8
             ─────
             1 2 0
             1 1 9
             ─────
               1 0 0
                 8 5
                 ─────
                 1 5 0
                 1 3 6
                 ─────
                   1 4 0
                   1 3 6
                   ─────
                     4 0
                     3 4
                     ─────
                       6 0
                       5 1
                       ─────
                         9 0
                         8 5
                         ─────
                           5 0
                           3 4
                           ─────
                           1 6 0
                           1 5 3
                           ─────
                             7 0
                             6 8
                             ─────
                               2
                               ─────
```

We now give a much shorter solution of this digital problem. We assume once more that the number we are trying to find is

$$N = a_n \, a_{n-1} \, a_{n-2} \, . \, . \, . \, . \, . \, a_2 \, a_1 \, a_0 \, ,$$

and that

$$3N/2 = a_{n-1} \, a_{n-2} \, . \, . \, . \, . \, . \, . \, . \, . \, a_2 \, a_1 \, a_0 \, a_n \, .$$

We consider the recurring decimal

$$x = a_n \, . \, a_{n-1} \, a_{n-2} \, . \, . \, . \, . \, a_2 \, a_1 \, a_0 \, \dot{a}_n \, a_{n-1} \, . \, . \, . \, a_1 \, \dot{a}_0 \, ,$$

where the recurring digits are precisely those which occur in N (the theory of recurring decimals will be described in Chapter VII). We now have

$$x/10 = . \, a_n \, a_{n-1} \, . \, . \, . \, . \, a_1 \, a_0 \, a_n \, a_{n-1} \, . \, . \, . \, a_1 \, a_0 \, . \, . \, . \, . \, . \, . \, ,$$

whereas

$$x - a_n = . \, a_{n-1} \, a_{n-2} \, . \, . \, . \, a_1 \, a_0 \, a_n \, a_{n-1} \, a_{n-2} \, . \, . \, . \, a_0 \, a_n \, . \, . \, . \, .$$

The first decimal, $x/10$, is a pure recurring decimal in which the digits which recur are those in N, and the second decimal $x - a_n$ is also a pure recurring decimal in which the digits which recur are those in $3N/2$. It follows that

$$x - a_n = \frac{3}{2} \frac{x}{10} \, ,$$

so that

$$17x = 20 \, a_n \, .$$

For the least value of the unknown integer N take $a_n = 1$, and then $x = 20/17$. If the reader has never worked out a decimal of this kind, he should do so. It is a pure recurring decimal, with a sequence of figures which recurs right from the beginning, after the decimal point. It is, in fact

$$x = 1 \, . \, 1764705882352941 \; 1764705882352941 \, . \, . \, . \, . \, . \, .$$

which is usually written

$$x = 1 \, . \, \dot{1}76470588235294\dot{1}.$$

If we now look back at the definition of x, we see that

$$N = 1, 176, 470, 588, 235, 294,$$

as found previously.

Both of these proofs have their points. The first is more elementary, but more tedious; the second uses the theory of infinite decimal expansions, and so overcomes the difficulty involved in the shift of an integer from the extreme left to the extreme right, by making the set of integers follow after one another in an infinite series. Each method will find its defenders, and yet the problem itself is one which most mathematicians would call *trivial*, because it leads nowhere, and does not fit into a general theory. The remainder of this chapter will be devoted to problems which are aspects of a general theory.

Numbers need not always be expressed in powers of ten. We shall see that it is often very useful to express them in powers of two, or powers of three. As we remarked at the beginning of this chapter, practically every integer has its devotees. This enthusiasm seems to fall off for numbers between 50 and 60, but starts up again at 666, the number of the Beast. On a more practical plane, it has been seriously suggested that we should take 12 as the basis of our number system. Multiplication tables become simpler, and fractions in constant use, like $\frac{1}{3}$, would be represented by the decimal 0.4 instead of by 0.333 . . . as in our present notation. It is certainly hard to justify the use of 10 as a base for numbers, with 12 pennies in a shilling, 20 shillings in a pound, 12 inches to a foot, 3 feet to a yard, 1,760 yards to a mile, and so on. A serious attempt was made in Victorian times to persuade Parliament that we should go over to a decimal currency. Readers of Trollope will remember that the Chancellor of the Exchequer who subsequently became Duke of Omnium was absorbed in the question. The attempt came to nothing. The arguments used against the change were not rational ones; but they were effective.

Perhaps the suggestion that 16 should take the place of 10 is worth mentioning. A book published in Philadelphia in 1862 seriously advocated a new number system, with the 16 fundamental numbers An, De, Ti, Go, Su, By, Ra, Me, Ni, Ko, Hu, Vy, La, Po, Fy, Ton, and then Ton-an, Ton-de etc. for 17, 18. . . As if this were not enough, the year was to be divided into sixteen months: Anuary, Debrian, Timander, Gostus, Suvenary, Bylian, Ratamber, Mesudius, Nictoary, Kolumbian, Husamber, Vyctorius, Lamboary, Polian, Fylander and Tonborius.

Mathematicians have often been thought of as queer people; but the author of the book just mentioned was a successful engineer. Another engineer, the Henry Archer who suggested the perforated border for postage-stamps, and the method of doing it, for which he received £4,000 reward, was a confirmed flat-earther, and was able to prove that the earth was not round from the Pyramids and some caves in Arabia. But perhaps we had better return to our numbers. . . .

The expression of numbers in the scale of 2, the *binary scale*, is of great theoretical importance. Instead of using powers of 10, we use powers of 2. If we divide a number by 2, the remainder is either 1 or 0. If the quotient is greater than 1, we may divide by 2 again, and find a remainder which is either 1 or 0. We continue the process until the quotient is 1. The results of this continued division enable us to express the given number as a sum of powers of 2. For instance

$$\tfrac{5}{2} = 2 + \tfrac{1}{2},$$
$$\tfrac{2}{2} = 1.$$

Hence $5 = 1 . 2^2 + 0 . 2 + 1,$

Similarly $6 = 1 . 2^2 + 1 . 2 + 0,$
$$7 = 1 . 2^2 + 1 . 2 + 1,$$
$$8 = 1 . 2^3 + 0 . 2^2 + 0 . 2 + 0,$$
$$9 = 1 . 2^3 + 0 . 2^2 + 0 . 2 + 1,$$

and so on. The integers multiplying the various powers of 2, called *coefficients*, are either 1 or 0.

If we merely write down the *coefficients* of the different powers of 2, and write

$$5 = 1 \ 0 \ 1,$$
$$6 = 1 \ 1 \ 0,$$
$$7 = 1 \ 1 \ 1,$$
$$8 = 1 \ 0 \ 0 \ 0,$$
$$9 = 1 \ 0 \ 0 \ 1,$$

we say that the numbers have been expressed as *binary decimals*. The only integers which appear in binary decimals are 0 and 1, if we allow 0 to be an integer. These binary decimals are exactly analogous to ordinary numbers, the position of each integer expressing the power of 2 which multiplies it, instead of the power of 10.

In calculations on modern electronic machines, numbers are usually expressed as binary decimals before being fed into the machine. Most electronic devices have two well-defined states, "on" or "off", "charged" or "not charged", and these are effectively represented by 0 and 1. Actual calculations in the binary scale are very simple. If, for example, we add 5 to 6, we add 1 0 1 to 1 1 0 in the binary scale, and obtain 2 1 1 at first. But since $2 . 2 = 2^2$, we must replace the 2 in the sum by 0, and insert 1 on the left, obtaining 1 0 1 1, which is the binary decimal for 11.

Similarly, if we multiply 5 and 6 in binary decimals we have

$$
\begin{array}{r}
1\ 0\ 1 \\
1\ 1\ 0 \\
\hline
1\ 0\ 1\ 0 \\
1\ 0\ 1 \\
\hline
1\ 1\ 1\ 1\ 0
\end{array}
$$

which is
$$2^4 + 2^3 + 2^2 + 2 + 0 = 30.$$

It is true that more figures are involved than in calculations in the scale of 10, but this is more than counterbalanced by the advantage of only having to deal with the simple operations expressed by

$$1 + 1 = 0 \text{ (put 1 in on the left)}, 1 + 0 = 1, 1 \cdot 1 = 1.$$

We now give an entertaining application of the theory of binary decimals. The following table of numbers is shown to the uninitiated person, and he is asked to think of a number between 1 and 63 inclusive. "Tell me", you say, "which columns of the table you can find the number in. That is, tell me the number at the head of each column, without telling me the number you are thinking of". When

1	2	4	8	16	32	1	2	4	8	16	32
1	2	4	8	16	32	33	34	36	40	48	48
3	3	5	9	17	33	35	35	37	41	49	49
5	6	6	10	18	34	37	38	38	42	50	50
7	7	7	11	19	35	39	39	39	43	51	51
9	10	12	12	20	36	41	42	44	44	52	52
11	11	13	13	21	37	43	43	45	45	53	53
13	14	14	14	22	38	45	46	46	46	54	54
15	15	15	15	23	39	47	47	47	47	55	55
17	18	20	24	24	40	49	50	52	56	56	56
19	19	21	25	25	41	51	51	53	57	57	57
21	22	22	26	26	42	53	54	54	58	58	58
23	23	23	27	27	43	55	55	55	59	59	59
25	26	28	28	28	44	57	58	60	60	60	60
27	27	29	29	29	45	59	59	61	61	61	61
29	30	30	30	30	46	61	62	62	62	62	62
31	31	31	31	31	47	63	63	63	63	63	63

he has done this, you can immediately tell him, without looking at the table, the number he had chosen.

For example, suppose that the number 51 was chosen. This occurs in the columns headed by the numbers 32, 16, 2 and 1. All that you need to do is to *add* these numbers to obtain the number 51.

The explanation is simple. The column headed 1 contains those numbers whose last digit in the binary decimal notation is 1. All odd numbers come in this column. They are of the form $2k + 1$. The column headed 2 contains all those numbers whose next-to-last digit in the binary decimal notation is 1. These numbers are of the form $4k + 2$, or $4k + 3$. For instance, the numbers 2, 3 ($k = 0$) and 6, 7 ($k = 1$) are in this column: and so on. The number 31 has the binary decimal 1 1 1 1 1, being equal to $2^4 + 2^3 + 2^2 + 2 + 1$, and is therefore in the columns headed 1, 2, 4, 8 and 16.

When the columns in which a given number is to be found are known, *the expression of the given number as a binary decimal is known*, and therefore the number itself is found by adding the appropriate powers of 2, which are given at the head of each column. To make the solution more mystifying, all that one need do is to number the columns 0, 1, 2, 3, 4 and 5, and then, when finding the number, to add 1 for column 0, 2 for column 1, 4 for column 2, 8 for column 3, 16 for column 4, and 32 for column 5, corresponding to the successive powers of 2;

$$2^0 = 1, \ 2^1 = 2, \ 2^2 = 4, \ 2^3 = 8, \ 2^4 = 16, \ 2^5 = 32.$$

The theory of binary decimals can be applied to the game of Nim, which we now describe. This game is played by two persons, and is most easily played with matches. Put a number of matches in three heaps, each heap containing an arbitrary number. Each player in turn takes at least one match from one, and only one of the heaps. He may take more than one, and there is no restriction on the number taken. The player who removes the very last match wins the game.

Let us take an example, in which the game starts with 2, 5 and 6 matches in the three heaps. Player A takes both matches from the first heap, leaving 0, 5, 6 matches in the three heaps. Player B then takes 5 matches from the third heap, leaving 0, 5, 1 matches in the heaps. If player A now takes 4 matches from the second heap, leaving 0, 1, 1 matches in the heaps, then player B is forced to take the one remaining match from either the second or third heap, and player A wins the game by taking the last remaining match from either the third or second heap.

The name Nim is supposed to be derived from the German instruction "Nehme eins", or "Take one", but, like all derivations, this has been disputed. In any case, the game is a good one, and the reader is invited to play it, and to work out a method of winning! It will soon become clear that all depends on the initial set-up of the game, the numbers of matches in the three heaps, and also on whose turn it is to start play. For example, with 0, 1, 1 matches in the three heaps, as above, the player who goes first must lose. We shall show how a real insight into the game can be obtained through the theory of binary decimals.

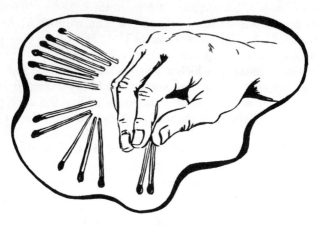

Fig. 1.

Because there is this mathematical theory, it has been possible to make an electronic machine which can play Nim against a human opponent. Such a machine was exhibited at the Festival of Britain Science Exhibition. The numbers in the three heaps were recorded by figures on three illuminated panels, and the removal of objects from one of the heaps was symbolised by rotating the appropriate knob through 1, 2, 3, . . divisions.

The humans playing against the "electronic brain" thought long and hard before each of their moves. The machine acted instantaneously when its turn came. At the end of a game, a panel flashed the announcement "Machine wins", or "Opponent wins". The author had the inestimable and unexpected privilege of being present with his two young sons when something went wrong. The human player had won, and stood looking rather pleased with himself; but the machine flashed "Machine wins"! It then changed its "mind", and flashed "Opponent wins"! Not stopping there, it went on to flash, with bewildering rapidity, first one message and then the other, until there was a rush of its several attendants behind the scenes, and the machine was switched off. When human beings become hysterical a sudden douche of cold water is often effective. In the case of an electronic instrument a sudden strong electric current around the faulty circuit sometimes has an equivalent effect. This was duly administered, and after this shock-treatment the faulty valve, amongst the many hundreds making up the apparatus, duly settled down, and there was no more trouble that day.

The mathematical analysis which follows will enable the reader to play at least as well as the electronic instrument, if not as quickly. We first give some definitions. We shall then explain the mathematical result: finally we shall prove it.

In the game of Nim the state of affairs at any moment is defined by the numbers of matches present in each of the piles. We can arbitrarily call one pile the first pile, another the second pile, and the final one the third pile, and then describe the state of affairs by giving the number of matches in each pile. If there be a matches in the first pile, b matches in the second pile and c matches in the third, the set of numbers (a, b, c) is a sufficient description of the state of affairs, and is called a *position* in the game. This position is said to be *held* by the player who has just taken one or more matches from a pile. It is called a *strategic position* if the player who holds it can force his opponent to lose, no matter how the opponent plays from then on. This means that if a player A occupies a strategic position, his opponent B cannot capture a strategic position on his next play; but on A's next play A can obtain a new strategic position, regardless of what B has done. Of course, strategic positions do not occur in all games. The interest of Nim is precisely that strategic positions do occur. In such a game the amateur, who plays by instinct, stands no chance against the professional, who knows what he is doing.

If we generalise the numerical example given above it is clear that $(0, n, n)$ is a strategic position, for any value of n. If A has just played, then whatever B does, A merely has to keep the number of matches in the two non-empty piles the same. As soon as B takes all the remaining matches from the second or third pile, A wins by taking all the remaining matches from the third or second pile. The reader will probably discover other strategic positions for himself. The theorem which tells us all we need to know is the following:

Let the numbers a, b, c of a position (a, b, c) be written in the binary scale. Then the position (a, b, c) is strategic if and only if the sum of the digits in each place is even.

For example, the position (5, 43, 46) is strategic, since

$$
\begin{array}{rll}
5 & = \text{binary} & \qquad\quad 1\ 0\ 1 \\
43 & = \text{binary} & 1\ 0\ 1\ 0\ 1\ 1 \\
46 & = \text{binary} & 1\ 0\ 1\ 1\ 1\ 0 \\
\hline
& & 2\ 0\ 2\ 2\ 2\ 2
\end{array}
$$

It must be pointed out that only the sums of individual columns are considered. There is no carry-over from one column to the next as there would be in ordinary addition of binary decimals. The reader need not follow the mathematical analysis which we now give if he can apply the rule. Take an example with smaller numbers, say the position (5, 8, 13). Then $5 = 2^2 + 1$, $8 = 2^3$, and $13 = 2^3 + 2^2 + 1$, so that the three binary decimals are

$$
\begin{array}{rl}
5 = & \quad 1\ 1 \\
8 = & 1\ 0\ 0 \\
13 = & 1\ 1\ 1 \\
\hline
& 2\ 2\ 2
\end{array}
$$

and since addition of columns gives an even sum in each column, the position is strategic. It will be seen that whatever move the next player makes, the sum of the columns will not be even. Suppose, in fact, that he takes 8 matches from the third heap. Then the three binary decimals are 1 1, 1 0 0, 1 1, and the columns add up to 1 2 2. All that the first player has to do is to make a move which will bring back the evenness of the columns. A simple way to do this is to take all the 8 matches in the second heap. We now have the position (5, 0, 5), and we have seen that this is strategic, and that the player who moves now cannot win. The mathematical discussion now follows. This merely states in precise terms what happens in a move from what we have called a strategic position.

If the position (a, b, c) is such that the sum of the digits in each place in the binary decimal representation is even, then any position which can be reached in one play from this position is not of this type.

For, by the rules of the game, just one of the piles, or numbers a, b, c is changed. Such a change produces a redistribution of the digits 0 and 1 in the binary decimal of the number changed. Therefore at least one 1 must become a 0, or one 0 must become a 1 in the binary decimal of the number which has changed. Hence in at least one column, the sum of the digits changes by 1; that is from an even number to an odd number.

We give another numerical verification of this theorem, using the (5, 43, 46) position, which, as we saw, is strategic. If 3 matches are taken from the third pile, so that 43 matches are left,

$$43 = \text{binary } 1\ 0\ 1\ 0\ 1\ 1$$

(observe the change of 1's to 0's and 0's to 1's), and the sum of the columns is now 2 0 2 1 2 3. There has been a change from even to odd in the fourth and sixth columns, corresponding to the change in the fourth and sixth columns of the binary decimal.

We now prove that if our opponent occupies a position in which the columns of the binary decimals do not all add up to even numbers, we can make a move which will ensure that they do. We scan the columns, and moving from the left we find the first column for which the sum of the digits is odd. This sum, since only three figures are added, and each is either 0 or 1, must be either 1 or 3. If the sum is 1, then just one of the three numbers being added has a 1 in this column, and this is the number which must be changed. If the sum is 3, each of the three numbers has a 1 in this column, and we change any one of them. Having decided which number is to be changed, replace the 1 in the binary decimal of this number by 0, and alter the digits which follow it by interchanging 1's and 0's in such a way that the new sums of digits are even in every column. It might be thought that this would *increase* the number, but since the first digit from the left which was altered was *reduced*, the new number is less than the original number, even if all the following digits were increased. Hence the change in the number is equivalent to a *subtraction*. Thus taking a match or matches from just one pile can alter an unstrategic position into what we have called a strategic position.

For example, when we have the position (5, 43, 43), and

$$
\begin{array}{rl}
5 = \text{binary} & 1\ 0\ 1 \\
4\ 3 = \text{binary} & 1\ 0\ 1\ 0\ 1\ 1 \\
4\ 3 = \text{binary} & 1\ 0\ 1\ 0\ 1\ 1 \\
\hline
& 2\ 0\ 2\ 1\ 2\ 3 \\
\hline
\end{array}
$$

the column to be changed is the fourth from the left. The integer 1 occurs in the binary decimal of the first pile, and we change this to 0. We then change the last digit so that the three digits in the last column add up to an even number. Since we must necessarily change this last digit to 0, we have removed *all* the matches from the first pile, and we have the strategic position (0, 43, 43).

To complete the proof of the first and fundamental theorem above we need only observe that the winning position (0, 0, 0), when all the piles are empty, is of the same special type as the positions we have called strategic. From what we have said above it is clear that once a player captures a position of this special type, he cannot be dislodged from it. It is equally clear that a sequence of such positions must ultimately lead to the winning position, since the number of matches in the piles diminishes at each play. Furthermore, these special positions are the only ones which can be maintained. Thus they are the only strategic positions.

As play proceeds one of the piles of matches must become empty. If the position is then a strategic position, it is easy to see that it must be of the type (0, n, n), with the two non-empty piles containing the same number of matches. Playing from this stage onwards is easy, but the game can be lost before this stage is reached! In general the initial position when play starts is not strategic, so that the player who begins has the advantage, since he can make his position strategic. Only if the initial position is strategic is the advantage with the second player, if he knows how to play. A list of strategic positions other than (0, n, n) is now given, for any game in which each pile has at most 15 matches in it.

1,	2,	3.	2,	4,	6.	3,	4,	7.
1,	4,	5.	2,	5,	7.	3,	5,	6.
1,	6,	7.	2,	8,	10.	3,	8,	11.
1,	8,	9.	2,	9,	11.	3,	9,	10.
1,	10,	11.	2,	12,	14.	3,	12,	15.
1,	12,	13.	2,	13,	15.	3,	13,	14.
1,	14,	15.						

4,	8,	12.	5,	8,	13.	6,	8,	14.	7,	8,	15.
4,	9,	13.	5,	9,	12.	6,	9,	15.	7,	9,	14.
4,	10,	14.	5,	10,	15.	6,	10,	12.	7,	10,	13.
			5,	11,	14.	6,	11,	13.	7,	11,	12.

If the reader can memorise these positions, he can be certain of winning if he plays Nim against men or machines. The complete mathematical analysis which can be given to games of this sort has led to an extensive study of "games" in recent times. They are of importance in war, where "games" are, of course, rather more complicated and deadly than the innocent game of Nim.

As a final example of the use which can be made of the binary representation of an integer, we shall show how to make a set of punched cards which should afford the reader a good deal of pleasure. Punched cards are used extensively in all kinds of offices. Suppose that an insurance company requires to know how many of its policyholders are over fifty. Each policy-holder has a card on which information is recorded by means of punched holes. There are machines which sort these cards into any desired category. If all the cards are put into the machine, and the machine set correctly, a batch of cards will soon fall out corresponding to all policy-holders over the age of fifty.

In a recent American film these machines were shown in action, in Washington D.C., of course, when it was desired to choose, from amongst millions of Americans, twelve men who might be able to guide a missile into the stratosphere with the object of capturing a meteorite before it burned away! The machines combed millions of cards for a combination of the qualities required from these supermen. We shall not describe these machines here, but give a simple example of the sorting of twelve punched cards into their correct order, once they have been disarranged, by means of the simplest possible automatic device, a pencil.

First obtain twelve cards! Those used in card-indexing are suitable. Mark the cards from 1 to 12. Now each of these numbers has a binary decimal representation, which we give below:

$$
\begin{array}{ll}
1 = 0\ 0\ 0\ 1 & \qquad 7 = 0\ 1\ 1\ 1 \\
2 = 0\ 0\ 1\ 0 & \qquad 8 = 1\ 0\ 0\ 0 \\
3 = 0\ 0\ 1\ 1 & \qquad 9 = 1\ 0\ 0\ 1 \\
4 = 0\ 1\ 0\ 0 & \qquad 10 = 1\ 0\ 1\ 0 \\
5 = 0\ 1\ 0\ 1 & \qquad 11 = 1\ 0\ 1\ 1 \\
6 = 0\ 1\ 1\ 0 & \qquad 12 = 1\ 1\ 0\ 0
\end{array}
$$

We punch holes in the top of each card to represent the binary decimal of the number marked on it. We punch a round hole for every 0, but for every 1 present in the binary decimal we make a slit reaching to the edge of the card. For example, on card 5 the holes would look like this:

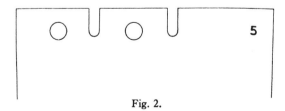

Fig. 2.

When all the cards have been punched, stack them in any order, with the punched holes at the top of each card. We now show how to put the cards in correct serial order, running from 1 to 12, without looking at the numbers on the individual cards. All we have to do is this: we stick a pencil through all the holes in the fourth position, on the extreme right, and gently raise the pencil. Those cards which have a circular hole in the fourth position, like cards number 2, 4, 6, 8, 10 and 12 will stay on the pencil. Those with a slit in the fourth position will drop off. Stack these, *in the order in which they drop off,* at the end of the pack. Then repeat the process, putting the pencil through all the holes in the third position, stacking the cards which drop off, in the order in which they drop off, at the back of the pack. The second hole is operated on next, and finally the first hole. When the cards which drop off are placed at the end of the pack, it will be seen that the cards are in the correct order, 1, 2, ... , 11, 12!

To show how the process works, suppose that the original order is

$$5, 3, 6, 9, 10, 12, 4, 2, 1, 11, 7, 8.$$

At the first operation the cards which drop off are, in order,

$$5, 3, 9, 1, 11, 7,$$

corresponding to the fact that each of these numbers has a 1 in the fourth place in its binary decimal representation, so that the corresponding card has a slit and not a hole in the fourth position. When these cards are stacked, in the order in which they drop off, at the end of the pack, the new order is

$$6, 10, 12, 4, 2, 8, 5, 3, 9, 1, 11, 7.$$

At the second operation, when the pencil is put through the holes in the third position, the cards

$$6, 10, 2, 3, 11, 7$$

drop off, since these numbers have a 1 in the third place of their binary representation. When these are stacked at the end of the pack the order becomes

$$12, 4, 8, 5, 9, 1, 6, 10, 2, 3, 11, 7.$$

After putting the pencil in the second hole from the extreme left, and stacking the cards

$$12, 4, 5, 6, 7$$

which drop off at the end, we have the order

$$8, 9, 1, 10, 2, 3, 11, 12, 4, 5, 6, 7.$$

When the pencil is put through the first hole, the cards

8, 9, 10, 11, 12

drop off, and when these are stacked at the end of the pack the order is

1, 2, 3, 4, 5, 6, 7, 8, 9, 10, 11, 12.

The reader will find that such a set of cards, which are easily made, will enable him to fulfil his social obligations at a party with the utmost ease.

We now come to a problem which is said to have been planted over here during the war by enemy agents, since Operational Research spent so many man-hours on its solution. This legend is probably as well founded as the one that Operational Research worked out the exact number of bombs necessary to subdue the island of Pantelleria, and it came to pass that when this number was dropped, lo and behold, the white flag was immediately flown! Whatever the truth of the legend, the problem is a fascinating one, and if the reader has not already come across it, he and his friends will spend many happy hours on its solution.

You are given 12 coins, pennies say, which are identical in appearance. One of these pennies, you are told, is defective, being either underweight or overweight, you are not told which. The other pennies are sound. The only weighing apparatus supplied is an equal-arm balance, but there are no weights. You are asked to find the defective penny in *three* weighings, and to say whether it is overweight or underweight.

Since there are no weights, all that can be done is to compare the weights of sets of pennies. If six of the pennies are weighed against the remaining six, one side of the scales will go down; but this may be for one of two reasons. Either the side which goes down contains a defective *heavier* coin, or the side which goes up contains a defective *lighter* coin. No information is therefore obtained from weighing six coins against the remaining six.

Progress is made if the coins are divided into three sets of 4, which we can call A, B and C. If we weigh A against B, and B against C, then either A and B will not balance, or B and C will not balance. For there is only one defective coin, and therefore two of the sets are perfect. If A and B balance, and also B and C, it would mean that each set of four coins has the same weight, which is not the case. If A and B balance, then the defective coin is in the set C, and if the set B is heavier than the set C, the defective coin is *lighter*. This kind of argument shows that *two* weighings will determine whether the defective coin is heavier or lighter, and pin it down, as one of a set of four. How can we find it in just one more weighing? If we could carry out *four* weighings, how easy it would be! But the question says it must be done in *three* weighings, and we shall show that it *can* be done. Those who are not professional mathematicians have a better record for solving this problem than the mathematicians, and it is hoped that the reader will take up the challenge before he reads further. The solution we shall give is a purely mathematical one, but it leads to a practical solution. A solution is possible which uses no mathematical notation, but a fair amount of accurate reasoning.

The solution we shall give uses the representation of numbers in the *ternary* scale, the scale of 3. If the reader has followed us so far, through the binary scale, the representation of numbers as sums of powers of 3 will not worry him unduly. The numbers which multiply the powers of 3, the *coefficients*, can be 0, 1 or 2 in this case. We give the ternary representation of the integers 1 to 12, since we shall be using this representation when we give the solution of the twelve-pennies problem:

$$1 = 0 \cdot 3^2 + 0 \cdot 3 + 1,$$
$$2 = 0 \cdot 3^2 + 0 \cdot 3 + 2,$$
$$3 = 0 \cdot 3^2 + 1 \cdot 3 + 0,$$
$$4 = 0 \cdot 3^2 + 1 \cdot 3 + 1,$$
$$5 = 0 \cdot 3^2 + 1 \cdot 3 + 2,$$
$$6 = 0 \cdot 3^2 + 2 \cdot 3 + 0,$$
$$7 = 0 \cdot 3^2 + 2 \cdot 3 + 1,$$
$$8 = 0 \cdot 3^2 + 2 \cdot 3 + 2,$$
$$9 = 1 \cdot 3^2 + 0 \cdot 3 + 0,$$
$$10 = 1 \cdot 3^2 + 0 \cdot 3 + 1,$$
$$11 = 1 \cdot 3^2 + 0 \cdot 3 + 2,$$
$$12 = 1 \cdot 3^2 + 1 \cdot 3 + 0.$$

We may therefore use the following ternary decimal notation for the integers between 1 and 12 inclusive:

1 = 0 0 1		7 = 0 2 1
2 = 0 0 2		8 = 0 2 2
3 = 0 1 0		9 = 1 0 0
4 = 0 1 1		10 = 1 0 1
5 = 0 1 2		11 = 1 0 2
6 = 0 2 0		12 = 1 1 0

Now for the twelve coin problem. We first stick labels on each coin, and number them from 1 to 12. Each label must be big enough to carry the number of the coin, its representation as a ternary decimal, and one other ternary decimal, which we shall now describe. In the ternary decimals given above, change every 0 into a 2, leave every 1 unaltered, and change every 2 into 0. Thus 0 0 2 becomes 2 2 0, and 0 1 0 becomes 2 1 2. On every label we write the ordinary ternary decimal which goes with the number on the label, and also the one obtained by this interchange of integers. Thus the penny marked 2 will contain the decimals 0 0 2 and also 2 2 0. Now, when we read out the expression of a number as a ternary decimal, beginning on the left, the first *change* of digit may be from 0 to 1, or from 1 to 2, or from 2 to 0, as in the cases 0 0 1, 0 1 1, 0 1 2, there being no examples within our range of the change from 2 to 0. If this is the case, we say that the decimal is a *clockwise* label for the number it represents. It is easy to verify that every coin has written on its label one clockwise decimal, and another, which we call *anticlockwise*. We shall mark the anticlockwise decimals with an asterisk, and then have the following decimals marked on the twelve labels:

$$
\begin{array}{rcl}
1 = 0\ 0\ 1 & = & 2\ 2\ 1^* \\
2 = 0\ 0\ 2^* & = & 2\ 2\ 0 \\
3 = 0\ 1\ 0 & = & 2\ 1\ 2^* \\
4 = 0\ 1\ 1 & = & 2\ 1\ 1^* \\
5 = 0\ 1\ 2 & = & 2\ 1\ 0^* \\
6 = 0\ 2\ 0^* & = & 2\ 0\ 2 \\
7 = 0\ 2\ 1^* & = & 2\ 0\ 1 \\
8 = 0\ 2\ 2^* & = & 2\ 0\ 0 \\
9 = 1\ 0\ 0^* & = & 1\ 2\ 2 \\
10 = 1\ 0\ 1^* & = & 1\ 2\ 1 \\
11 = 1\ 0\ 2^* & = & 1\ 2\ 0 \\
12 = 1\ 1\ 0^* & = & 1\ 1\ 2.
\end{array}
$$

The point that is fundamental to the rest of the discussion is that, given any one of the 24 decimals listed above, we can identify the coin to which it belongs, since every coin has a unique (clockwise or anticlockwise) ternary decimal associated with it, and every clockwise or anticlockwise decimal determines uniquely the corresponding anticlockwise or clockwise decimal which is associated with the same coin. If, for example, the ternary decimal 2 0 1 were to appear during the subsequent discussion, we should see at once that it is a clockwise label, and associated with the anticlockwise label 0 2 1, which represents the number 7; and so on.

Now that every coin is labelled, we can begin the weighings. We wish to detect the false coin amongst the twelve in just three weighings, and to be able to say whether it is lighter or heavier than the true coins. The method used makes the false coin *write its own signature* in the ternary decimal notation. For each of the three weighings we shall divide the twelve coins into three groups of four.

First weighing: We put into the first group, which we call $C(1, 0)$, the coins which have 0 for the first digit on their clockwise label. Hence

$C(1, 0)$ contains the coins 1, 3, 4 and 5.

In the second group, which we call $C(1, 1)$, we put the coins which have 1 as the first digit on their clockwise label. Hence

$C(1, 1)$ contains the coins 9, 10, 11 and 12.

The remaining group $C(1, 2)$ contains the coins 2, 6, 7 and 8, which have 2 for the first digit of their clockwise labels.

We now begin weighing, putting the group $C(1, 0)$ in the left-hand pan, and the group $C(1, 2)$ in the right-hand pan. We mark up the results of the weighings on a scoring-board. If the right-hand pan goes down, we mark up the integer 2. If the left-hand pan goes down, we mark up the integer 0. If the scale-pans remain level, we mark up the integer 1. *We now have the first digit in the label which will identify the defective coin.* The remaining two weighings are similar in character, so that if this first one is understood, the rest follows.

Second weighing: We redivide the twelve coins up into three groups according to the integer which appears in the *second* place on their clockwise labels.

$C(2, 0)$ contains the coins 1, 6, 7 and 8;
$C(2, 1)$ . . . . 3, 4, 5 and 12;
$C(2, 2)$ . . . . 2, 9, 10 and 11.

We weigh the first group in the left-hand pan against the third group in the right-hand pan. Once again, if the left-hand pan goes down we mark up 0 in the *second* position from the left on the scoring-board. If the right-hand pan goes down, we mark up 2, and if the pans balance we mark up 1. *We now have the first two integers in the label which will identify the faulty coin.*

Third weighing: We finally divide up the coins into three sets of four according to the digit which appears in the *third* place on their clockwise labels.

$C(3, 0)$ contains the coins 2, 3, 8 and 11;
$C(3, 1)$ . . . . 1, 4, 7 and 10;
$C(3, 2)$ . . . . 5, 6, 9 and 12.

Once again we put $C(3, 0)$ in the left-hand pan, and $C(3, 2)$ in the right-hand pan, marking up 0 in the third position on the scoring-board if the left-hand pan goes down, 2 if the right-hand pan goes down, and 1 if both pans balance. *We now have the final integer in the label which identifies the faulty coin.*

If the label marked up on the scoring-board is a clockwise label, the coin is overweight. If it is an anticlockwise label, the coin is under-weight. As an example, let us suppose that the coin marked 7 is the defective one, and is overweight.

In the first weighing the coin marked 7 is in $C(1, 2)$, and in the right-hand pan. This goes down, and so the first digit marked up on the board is 2.

In the second weighing the coin marked 7 is in the set $C(2, 0)$ and is therefore placed in the left-hand pan. This goes down, and there-fore the second integer marked up on the board is 0.

In the final weighing the coin marked 7 is in $C(3, 1)$, and is not in either pan. These therefore balance, and the third digit marked up is 1.

We therefore find the three digits 2 0 1 marked up on the board. This is a clockwise label, associated with the anticlockwise label 0 2 1, which is the ternary decimal for the digit 7.†

We complete our discussion of ternary decimals with a solution of a much simpler problem. Given an equi-arm balance, and the neces-sity of weighing between 1 and 40 pounds, how can we do this with the fewest number of weights? We show that, provided we can put the weights in either scale-pan, it is sufficient to have weights of 1, 3, 9 and 27 pounds. This is a deduction from the fact that in the ternary decimal representation, instead of using 0, 1 and 2 as co-efficients we may use 0, 1 and -1, provided that we alter the decimal appropriately. For example

$$11 = 1 \cdot 3^2 + 0 \cdot 3 + 2$$

can be written

$$11 = 1 \cdot 3^2 + 1 \cdot 3 - 1.$$

Again

$$8 = 0 \cdot 3^2 + 2 \cdot 3 + 2$$

can be written in the form

$$8 = 1 \cdot 3^2 + 0 \cdot 3 - 1.$$

It is clear that, in order to change the coefficient 2 into a coefficient -1, all that we need to do is to increase the integer coefficient on the

† This solution is due to F. J. Dyson and R. C. Lyness (*Math. Gazette*, Vol. 30, Oct. 1946).

left of the 2, in the ternary decimal representation, by 1. If this integer coefficient becomes 3, we must, of course, replace it by 0, and increase the coefficient on *its* left by 1. Now

$$40 = 1 \cdot 3^3 + 1 \cdot 3^2 + 1 \cdot 3 + 1,$$

and all numbers between 1 and 40 can be represented by a ternary decimal with four digits, these digits being either 0, 1 or -1. For example

$$38 = 1 \cdot 3^3 + 1 \cdot 3^2 + 0 \cdot 3 + 2$$
$$= 1 \cdot 3^3 + 1 \cdot 3^2 + 1 \cdot 3 - 1,$$

and so on. It follows that if we are given an equi-arm balance, we can weigh up to 40 pounds if we merely have weights of

$$1, 3, 3^2 \text{ and } 3^3$$

pounds, it being assumed that we can place these in either scalepan. To weigh 38 pounds, for instance, the above representation of 38 shows that we merely have to put 27 pounds, 9 pounds and 3 pounds in one pan, and 1 pound in the other. Similarly for any other weight between 1 and 40 pounds inclusive.

In bringing this chapter to an end, the author is conscious of the enormous amount of material on the ordinary integers which he has omitted. But there are many books which deal with this fascinating subject. Here an attempt has been made to show how mathematicians deal with a number of problems which cluster around the theory of the representation of integers as binary and ternary decimals. Very little has been said about prime numbers, whose properties have intrigued many for thousands of years. But since they have been mentioned, we can hardly illustrate the mathematician's art better than by giving the classical proof that *there is an infinite number of prime numbers*. We shall follow this with the proof that *the square root of 2 is not a rational number*, explaining our terms when we give the proof, and then go on to something quite different, the theory of probability, or choice and chance.

The *prime numbers* or *primes* are the numbers

$$2, 3, 5, 7, 11, 13, 17, 19, 23, 29, \ldots \ldots$$

which cannot be resolved into smaller factors. Thus 37 and 317 are prime. The primes are the material out of which all numbers are built up by multiplication: thus $666 = 2 \cdot 3 \cdot 3 \cdot 37$. Every number which is not prime itself is divisible by at least one prime, usually, of course, by several. We have to prove that there are infinitely many

primes, that is that the set of numbers above never comes to an end.
Let us suppose that it does, and that

$$2, 3, 5, \ldots \ldots , P$$

is the complete series, so that P is the largest prime. On this hypothesis we consider the number

$$Q = (2 . 3 . 5 \ldots \ldots P) + 1.$$

It is plain that Q is not divisible by any of 2, 3, 5, ... , P; for it leaves the remainder 1 when divided by any one of these numbers. But, if not itself prime, it is divisible by *some* prime, and therefore there is a prime (which may be Q itself) greater than any of them. This contradicts our hypothesis, that there is no prime greater than P; and therefore this hypothesis is false.

This proof goes back to Euclid, and is by *reductio ad absurdum*. This is one of the mathematician's finest weapons, and is used frequently.

The second theorem and proof go back to Pythagoras. If x is the square root of 2, we imply that $x^2 = 2$. A *rational number* is a fraction a/b, where a and b are integers. We assume that a and b have no common factor, since if they had, we could remove it. To prove that the square root of 2 is not a rational number, we have to show that the equation

$$(a/b)^2 = 2,$$
or
$$a^2 = 2 b^2$$

cannot be satisfied by integral values of a and b which have no common factor. This is a theorem of pure arithmetic. We argue again by *reductio ad absurdum*. We suppose that the last equation is true, a and b being integers without any common factor. It follows from this equation that a^2 is even (since $2b^2$ is divisible by 2), and therefore that a is even, since the square of an odd number is odd. If a is even, then

$$a = 2c,$$

for some integral value of c; and therefore

$$2b^2 = a^2 = (2c)^2 = 4c^2,$$
or
$$b^2 = 2c^2.$$

Hence b^2 is even, and therefore, for the same reason as before, b is even. That is to say both a and b are even, and therefore have the common factor 2. This contradicts our hypothesis, and therefore the hypothesis is false.

The late Professor G. H. Hardy said of these two theorems: "They are 'simple' theorems, simple both in idea and in execution, but there is no doubt at all about their being theorems of the highest class. Each is as fresh and significant as when it was discovered—two thousand years have not written a wrinkle on either of them."

CHAPTER II

CHANCE AND CHOICE

IF a coin is spun in the air, and alights on a flat inelastic surface, it will come to rest on one side or the other, and will therefore be either "heads" or tails". This is a matter of experience. But if we think about it, we see that there is also the possibility that the coin may come to rest on its edge because, like Buridan's ass, it may be unable to make a choice, and will therefore not know on which side to fall. In practice, of course, this does not happen.

Here we immediately run up against a fundamental difficulty in the theory of probability—its connection with the real physical world. What happens if we keep on tossing a coin? How many heads do we get, and how many tails? We feel that there is an equal chance of the coin coming down heads or tails, and that *on the average* the number of heads will be nearly the same as the number of tails, if the coin is spun a large number of times.

Questions relating to chance, in this era of football-pools, interest everybody. During the war a B.B.C. Brains Trust was asked: "What is the law of averages?" Without the slightest hesitation, and without even a provisional "It all depends on what you mean by 'law' or 'average'," the late Dr. C. E. M. Joad answered: "The law of averages says that if you spin a coin one hundred times, it will come down heads fifty times, and tails fifty times." The oracle had spoken. There was no discussion.

If the reader feels that this is a correct statement, he should spin a coin one hundred times. If he does get fifty heads and fifty tails exactly, the experiment is worth repeating. If he continues to get exactly fifty heads and fifty tails, the law of averages will not be working in his case! In fact, the law of averages is not a law, but expresses the feeling that unusual events do not occur frequently! It is unusual for a large family to consist entirely of girls, and if friends of ours have five daughters already, we assure them that the next child is bound to be a boy, by the law of averages! But this reasoning seems to be invalid in the case of the Dutch Royal family. We shall return to this point later. A few words about the historical background of the mathematics of choice and chance will set the stage for this chapter.

In 1654 the Chevalier de Méré proposed to Blaise Pascal a problem concerning the division of stakes in a game of dice. Pascal corresponded with Fermat on this and related problems, and it is clear that the modern theory of probability first saw the light of day in these discussions. Pascal and Fermat were concerned with what is called in English "the problem of points": two players each want a certain number of points in order to win a game. If they separate without playing out the game, how should the stakes be divided between them? The question amounts to asking what are the chances (or what is the probability) which each player has, at any given stage of the game, of winning the game. In the discussion between Pascal and Fermat it was supposed that the players have equal chances of winning a single point.

Before we discuss the meaning of the term "probability" it is interesting to see that, writing to Fermat about de Méré, who had first consulted Pascal about these problems, Pascal says: "Il a très-bon esprit, mais il n'est pas géomètre: c'est comme vous savez, un grand défaut."* Pascal then gives reasons why, although de Méré is very intelligent, he cannot be considered a mathematician (the word *geometer* in those days, and for many years afterwards, was always used where we should say *mathematician*). Apparently de Méré could not see that a mathematical line is infinitely subdivisible, but persisted in believing that it only contains a finite number of points. Pascal could not get him over this difficulty, and naturally felt that this was a great defect.

But de Méré was not content with merely disagreeing with the great Pascal. From his denial of the infinite divisibility of a mathematical line, de Méré went on to deduce contradictions in the laws of arithmetic! This perseverance marks him out as a true mathematician. Although it was taken for granted up to the nineteenth century that the mathematics we know is free from contradiction, this has only been *proved* to date for the simplest operations. We shall return to this topic in Chapter VIII.

Since a spun coin can come down heads or tails, and since both possibilities are equally likely if the coin is not weighted on one side, and if the person spinning the coin does so in a random manner (we shall discuss what *random* means later on), we say that the chances of a coin coming down heads are 1 in 2, or that the *probability* of the coin coming down heads is $\frac{1}{2}$. If we wish to blend this explanation of the probability of a spun coin coming down heads being $\frac{1}{2}$ with the

---

* "He is very intelligent, but he is not a mathematician: this, as you know, is a great defect."

results of experience, we say that if a coin is spun *a large number* N of times, it will come down heads ($\frac{1}{2}$) N times.

Our use of the phrase "large number" is technical, and will be explained in Chapter VII. We therefore make another attempt: If a coin is spun N times, and if N' be the number of times it comes down heads, then the ratio N'/N will approach to the value $\frac{1}{2}$ as N increases indefinitely (that is, without bound). This last definition is called *the frequency definition* of probability.

If, in an actual experiment, the ratio N'/N did not approach $\frac{1}{2}$ as the number N of tosses increased without limit, we should say:

(1) the coin is not a true one; or

(2) the coin is not being spun in a random manner!

In fact, all that we can say of the adjective "random" is that it is used to describe behaviour which fits in with the mathematical theory of probability! Human beings cannot perform in a random manner. It is quite simple to make a coin-tossing machine which will always make a given coin fall on a given non-elastic surface with heads up if the coin is inserted in the machine in the same way. What should a man tossing a coin do to ensure that he is not behaving like this machine? How can he be sure that he is behaving in a random manner?

In agricultural experiments, which are based on probability theory, which itself postulates random behaviour, it is important to scatter seed over a given plot of ground in a random manner, so that no part of the plot shall be especially favoured. The customary procedure (when the author, as an undergraduate, was shown over the Cambridge University Farm) was for the experimenter to take a handful of seed, to shut his eyes, whirl madly round near the plot, and then to open his hand and to let go of the seed.

From the practical point of view there is little to object to in modern agricultural experiments, but the theoretical difficulties inherent in the definitions of probability have not yet been resolved. The importance of the subject is very great, since statistical theory is based on it, and this is largely used in the arts of both peace and war.

It is to be hoped that this preliminary discussion has not confused the reader, by giving him too much to think about in a lump. We shall adopt the following definition, having explained that some of the terms involved in the definition cannot be defined except by the use of a circular argument:

Let the number of ways in which a certain event can happen be h, and the number of ways in which it can fail to happen be f. Let us further suppose that the ways in which the event can happen or fail to happen are all equally likely (!). Then the probability that the event will happen is $h/(h + f)$, and the probability that the event will fail to happen is $f/(h + f)$.

The probability that the event will happen *or* fail to happen is

$$h/(h + f) + f/(h + f) = 1.$$

Since this is *certain*, we see that the probability of an event is a number lying between 0 and 1, and *certainty* is given the probability 1.

Fig. 3.

Examples in probability are usually concerned with spinning coins, throwing dice, or drawing coloured balls out of an urn. As the early history of the subject was bound up with gambling for several hundred years, dice, coins and packs of cards inevitably figure largely in the literature. Since urns and coloured balls are more aseptic, perhaps, we illustrate our definition of probability by considering an urn which contains 3 red and 7 white balls.

A ball can be drawn out in 10 ways, all equally probable, and of these drawings 3 will be of red, and 7 of white balls. Hence the probability of drawing a red ball is $\frac{3}{10}$, and that of drawing a white ball is $\frac{7}{10}$.

But dice have certain obvious advantages over urns, and cannot be ignored if we wish to illustrate some of the fundamental mathematical laws of probability. If we make one throw of a single die, what is the probability of throwing a 2?

Since the die has six faces, any one of which may turn up (if the die is not loaded), there is a total of six equally likely ways in which the desired event can occur or fail to occur. There is only one way in which the event can occur. Therefore the probability of throwing a 2 with a single throw of a die is $\frac{1}{6}$.

What is the probability of throwing a 2 *or* a 3 with one throw of a single die? Again there is a total of six ways in which the proposed event can occur or fail to occur. There are two ways in which it can occur. Therefore the probability is $\frac{2}{6}$, or $\frac{1}{3}$.

The same result can be arrived at in another way. Noting that the probability of a 2 being thrown is $\frac{1}{6}$, and that the probability of a 3 being thrown is $\frac{1}{6}$, we can argue that the probability of a 2 *or* a 3 being thrown is $\frac{1}{6} + \frac{1}{6} = \frac{1}{3}$, since the two events are *mutually exclusive*. This simple example illustrates the following general law:

Addition law of probabilities: If $p_1, p_2, \ldots, p_n$ are the respective probabilities of n mutually exclusive events, then the probability that *one* of the events will occur is the *sum* of these probabilities, which is

$$p_1 + p_2 + \ldots + p_n .$$

The reader need not worry much about this law if he understands what *mutually exclusive events* are. The proof we indicate underlines the meaning. We flit back to the frequency definition of probability, and consider any large number N of occasions where all the events are in question. Out of these N occasions the given n events will happen on

$$p_1 N, p_2 N, \ldots , p_n N$$

occasions respectively. There is no cross-classification necessary here, since the events are mutually exclusive, and therefore not more than one of the events can happen on any one occasion. Out of the N occasions, therefore, one or other of the events will happen on

$$p_1 N + p_2 N + \ldots + p_n N = (p_1 + p_2 + \ldots + p_n) N$$

occasions. Hence the probability that one out of the n events will happen on any one occasion is $p_1 + p_2 + \ldots + p_n$.

As another simple example of this law, we ask for the probability that in a single throw of a die we do not score more than 5. To satisfy the conditions, the die may show 1 or 2 or 3 or 4 or 5. Since these events are mutually exclusive, and the probability of each is $\frac{1}{6}$, the probability of one or other of these events is $\frac{5}{6}$. Of course, we can also argue that the probability of 6 *not* turning up is $1 - \frac{1}{6} = \frac{5}{6}$.

There is only one further law we wish to mention, the *multiplication law* of probabilities, which we first illustrate with an example. What is the probability of throwing two sixes with a pair of dice?

Since every number on the first die can be associated with six numbers on the second die, and since there are six numbers on the first die, the dice can fall in any one of 6 . 6 = 36 possible ways, as is illustrated below.

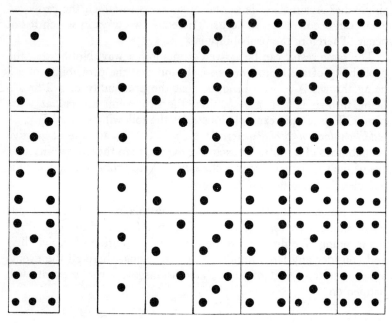

Fig. 4.

Of these 36 possibilities, only one is favourable—that in which a 6 turns up on each die. Hence the probability of throwing a double six is $\frac{1}{36}$. We note that the probability of throwing a 6 with the first die is $\frac{1}{6}$, and that the throwing of a 6 with the second die is independent of this, the probability also being $\frac{1}{6}$, and that $\frac{1}{36} = (\frac{1}{6})(\frac{1}{6})$. This illustrates the

Multiplication law of probabilities: If $p_1, p_2, \ldots, p_n$ are the probabilities of n independent events, then the probability that all n of the events will occur at once is the product of these probabilities, which is

$$p_1 p_2 \cdot \cdots \cdot p_n.$$

Again a proof is instructive. Out of any large number N of cases where all the events are in question, the first event will happen on p_1N occasions. Out of these p_1N occasions (we assume that p_1N is large with N) the second event will also happen on $p_2(p_1N)$ occasions. Hence there are p_1p_2N occasions out of N on which both the first and second events occur. Continuing in this way, we see that out of N occasions there are $p_1p_2 \ldots p_nN$ occasions on which *all* the events occur. The probability that all the events occur on any one occasion is therefore $p_1p_2 \ldots p_n$.

To illustrate the second law again, let us suppose that a die is thrown twice, and let us calculate the probability that the first throw does not exceed 3, and the second throw does not exceed 5. From the first law we see that the probability that the first throw does not exceed 3 is $\frac{3}{6}$, and the probability that the second throw does not exceed 5 is $\frac{5}{6}$. Now the result of the first throw in no way affects the result of the second throw. Hence, by the multiplication law the probability that both events take place, that is that the first throw will not exceed 3 and the second throw will not exceed 5, is $(\frac{3}{6})(\frac{5}{6}) = \frac{5}{12}$.

Although this book is intended to illustrate the art of the mathematician, we should not be painting a true picture if we omitted mention of some grievous errors into which great mathematicians, like ordinary mortals, have sometimes fallen. We might even say that without some *chiaroscuro* the picture would be an uninteresting one. D'Alembert, one of the great French mathematicians of the eighteenth century, slipped up over the following simple problem:

In two tosses of a single coin, what is the probability that heads will appear at least once?

The probability that tails turns up first toss is $\frac{1}{2}$, and the probability that it turns up second toss is also $\frac{1}{2}$. Hence, since these events are independent, the second law says that the probability that tails turn up on both tosses is $(\frac{1}{2})(\frac{1}{2}) = \frac{1}{4}$. The probability that this is *not* so is $1 - \frac{1}{4} = \frac{3}{4}$. But this is the probability that heads will appear *at least once* in both tosses. This is the correct answer. To follow D'Alembert's argument, and to see where he went wrong, we enumerate the possibilities below.

Noting that heads on the first toss can be associated with either heads or tails on the second toss, and that tails on the first toss can similarly be associated with either heads or tails on the second toss, the total number of possible cases is four. Of these four cases, the first three are favourable in that they contain at least one head. Therefore, once more, the required probability of obtaining at least one head in two tosses is $\frac{3}{4}$.

But when this problem was proposed to d'Alembert in 1754, he argued as follows: There are only three cases: heads on the first throw, or heads on the second throw, or heads not at all. Two of these cases are favourable—the first two. Therefore the desired probability is $\frac{2}{3}$, there being three cases in all!

It is not too difficult to see where this solution is wrong. D'Alembert's first case included our first *two* cases. In other words, "Heads on the first throw" meant, to d'Alembert, "heads on the first throw regardless of what happens on the second throw," whereas "heads on the second throw" meant "tails on the first throw followed by heads on the second throw."

It is true that one of the three possibilities in d'Alembert's system must occur, and that the possibilities are mutually exclusive. The trouble is that they are *not equally likely*. It is evident that "heads on the first throw regardless of what happens on the second throw" (our first two cases) is twice as likely as "tails on the first throw and heads on the second".

As another example of the possibility of error in problems on probability we consider two urns A and B, and suppose that A contains 3 white and 4 black balls, and that B contains 4 white and 3 black balls. A person selects one of the urns at random, and draws a ball. Find the chance that it may be white.

We might, with some plausibility, reason thus: The drawer must select either A or B. If he selects A, the chance of drawing white is $\frac{3}{7}$. If he selects B, the chance of drawing white is $\frac{4}{7}$. Hence, by the addition rule, the whole chance of drawing white is

$$\tfrac{3}{7} + \tfrac{4}{7} = 1.$$

In other words, white is *certain* to be drawn, which is absurd.

The mistake occurs in not taking account of the fact that the drawer has a choice of urns, and that the chance of his selecting A must be multiplied by his chance of drawing white *after* he has selected A. The probability of drawing white from A is therefore

$$(\tfrac{1}{2})(\tfrac{3}{7}) = \tfrac{3}{14}.$$

In the same way, the probability of drawing white from B is

$$(\tfrac{1}{2})(\tfrac{4}{7}) = \tfrac{4}{14},$$

so that the total probability of drawing a white ball is

$$\tfrac{3}{14} + \tfrac{4}{14} = \tfrac{1}{2}.$$

The necessity for introducing the factor $\frac{1}{2}$ can best be seen by reasoning directly from the fundamental notions. Let us suppose the drawer to make the experiment any large number N of times. In the long run the one urn will be selected as often as the other. Hence, out of N times, A will be selected $N/2$ times. Out of these $N/2$ occasions, white will be drawn from A

$$(\tfrac{3}{7})(N/2) = 3N/14$$

times. Similarly we see that white will be drawn from B a total of

$$(\tfrac{4}{7})(N/2) = 4N/14$$

times. Hence, on the whole, out of N trials white will be drawn

$$(\tfrac{3}{14} + \tfrac{4}{14})\, N$$

times. The probability of drawing white is therefore $\tfrac{1}{2}$.

The reader is perhaps acquiring a fuller realisation of what Eliza Doolittle, the heroine of Shaw's *Pygmalion*, meant when she answered "Not bloody likely!", when asked whether she would take a walk across the Park. We give some more examples to illustrate probability theory.

How many times must a man be allowed to toss a penny in order that the odds may be 100 to 1 that he gets at least one head?

Let n be the number of tosses. The complementary result to "at least one head" is "all tails". Since the chance of a tail each time is $\tfrac{1}{2}$, and the result of each toss is independent of the result of any other, the chance of "all tails" in n tosses is, by the multiplication law, equal to $(\tfrac{1}{2})^n$. The chance of one head at least in n tosses is therefore $1 - (\tfrac{1}{2})^n$. By the conditions of the question we must therefore have the equation

$$1 - (\tfrac{1}{2})^n = \tfrac{100}{101},$$

which leads immediately to

$$2^n = 101.$$

No elaborate mathematics is needed to find n. We know that $2^3 = 8$, so that $2^6 = 64$, and therefore $2^7 = 128$. Hence, if n were not an integer, n would lie between 6 and 7. We can say that in 6 tosses the odds are less than 100 to 1, but in 7 tosses the odds are greater.

Our next example is of great historical interest, since it was proposed by the diarist Samuel Pepys to Isaac Newton.

One man asserts that he will throw at least one six with six dice. A second man asserts that he will throw at least two sixes by throwing twelve dice. A third man asserts that he will throw at least three sixes by throwing eighteen dice. Which of the three men stands the best chance of carrying out his promise?

We shall give a solution of this problem, and then reproduce Newton's two letters to Pepys. We know that the probability of a die showing a six when thrown is $\frac{1}{6}$. The probability of the die *not* showing the six is $\frac{5}{6}$. Evidently

$$1 = (\tfrac{1}{6} + \tfrac{5}{6}) = \tfrac{1}{6} + \tfrac{5}{6},$$

the two terms on the right showing the probability of throwing a six, and the probability of not throwing a six when one die is thrown. We have written this obvious equation because it suggests the next step:

If we expand

$$1 = (\tfrac{1}{6} + \tfrac{5}{6})^2 = (\tfrac{1}{6})^2 + 2(\tfrac{1}{6})(\tfrac{5}{6}) + (\tfrac{5}{6})^2,$$

we see that the terms on the right give:

$(\tfrac{1}{6})^2$ = the probability of two sixes if two dice are thrown:

$2(\tfrac{1}{6})(\tfrac{5}{6})$ = the probability of just one six if two dice are thrown:

$(\tfrac{5}{6})^2$ = the probability of no sixes if two dice are thrown.

If we continue in this way, making use of the Binomial Theorem (see Chapter VI), we have

$$1 = (\tfrac{1}{6} + \tfrac{5}{6})^3 = (\tfrac{1}{6})^3 + 3(\tfrac{1}{6})^2(\tfrac{5}{6}) + 3(\tfrac{1}{6})(\tfrac{5}{6})^2 + (\tfrac{5}{6})^3,$$

and the terms on the right give the respective probabilities of throwing three sixes, two sixes, one six and no sixes when *three* dice are thrown.

The method is a general one, and when we throw six dice we consider the expansion

$$1 = (\tfrac{1}{6} + \tfrac{5}{6})^6 = (\tfrac{1}{6})^6 + 6(\tfrac{1}{6})^5(\tfrac{5}{6}) + \ldots + 6(\tfrac{1}{6})(\tfrac{5}{6})^5 + (\tfrac{5}{6})^6,$$

and the various terms on the right give the probabilities of throwing six sixes, five sixes, four sixes, three sixes, two sixes, one six and no six when six dice are thrown. To find the probability of *at least* one six being thrown, we add the probabilities that one, two, ... , six sixes are thrown. This sum is equal to

$$1 - (\tfrac{5}{6})^6 = \frac{(6)^6 - (5)^6}{(6)^6} = \frac{31031}{46656}.$$

To find the probability of throwing at least two sixes with twelve dice, we must consider the expansion

$$1 = (\tfrac{1}{6} + \tfrac{5}{6})^{12} = (\tfrac{1}{6})^{12} + 12(\tfrac{1}{6})^{11}(\tfrac{5}{6}) + \ldots + 12(\tfrac{1}{6})(\tfrac{5}{6})^{11} + (\tfrac{5}{6})^{12},$$

and we want the sum of all the terms on the right except the last two terms, so as to add the probabilities of two, three, ... , eleven, twelve sixes turning up. This sum is

$$1 - 12(\tfrac{1}{6})(\tfrac{5}{6})^{11} - (\tfrac{5}{6})^{12}$$

$$= \frac{(6)^{12} - 12(5)^{11} - (5)^{12}}{(6)^{12}}$$

$$= \frac{1,\,346,\,704,\,211}{2,\,176,\,782,\,336}\,\cdot$$

This probability is less than the first, and in a similar way we find that the probability of throwing three sixes at least with a throw of eighteen dice is even less than this.

In the first letter from Newton reproduced below, we see that Newton disentangles the *mathematical* problem from the one presented to him. This first step is often the most difficult one. In the second letter he obtains the above results by equivalent methods. We now reproduce the two letters from Newton to Pepys. They are fine examples of the working of a great genius.

Isaac Newton to S. Pepys.

Cambridge, November 26, 1693.

Sir—I was very glad to hear of your good health by Mr. Smith, and to have any opportunity given me of showing how ready I should be to serve you or your friends upon any occasion, and wish that something of greater moment would give me a new opportunity of doing it, so as to become more useful to you than in solving only a mathematical question. In reading the question, it seemed to me at first to be ill stated; and in examining Mr. Smith about the meaning of some phrases in it, he put the case of the question the same as if A played with six dice till he threw a six; and then B threw as often with twelve, and C with eighteen, the one for twice as many, the other for thrice as many, sixes. To examine who had the advantage, I took the case of A throwing with one dice, and B with two—the former till he threw a six, the latter as often for two sixes; and found that A had the advantage. But whether A will have the advantage when he throws with six, and B with twelve dice, I cannot tell; for the number of dice may alter the proportion of the chances considerably, and I did not compute it in this case, the problem being a very hard one. And, indeed, upon reading the question anew, I found that these cases do not come within the question; for here an advantage is given to A by his throwing first till he throws a six: whereas, the question requires, that they throw upon equal luck, and by consequence that no advantage be given to any one by throwing first. The question is this: A has six dice in a box, with which he is to fling a six; B has in another box twelve dice, with which he is to fling two sixes; C has in another box eighteen dice, with which he is to fling three sixes. Qy, whether B and C have not as easy a task as A at even luck? If this last question must be understood according to the plainest sense of the words, I think that sense must be this:

1st. Because A, B, and C, are to throw upon even luck, there must be no advantage of luck given to any of them by throwing first or last, by making anything depend upon the throw of any one, which does not equally depend on the throws of the other two: and, therefore, to bar all inequality of luck on these accounts, I would understand the question as if A, B, and C, were to throw all at the same time.

2ndly. I take the most proper and obvious meaning of the words of the question to be, that when A flings more sixes than one, he flings a six, as well as when he flings but a single six, and so gains his expectation: and so, when B flings more sixes than two, and C more than three, they gain their expectations. But if B throw under two sixes, and C under three, they miss their expectations; because, in the question, 'tis expressed that B is to throw two, and C three sixes.

3rdly. Because each man has his dice in a box, ready to throw, and the question is put upon the chances of that throw, without naming any more throws than that, I take the question to be the same as if it had been put thus upon single throws.

What is the expectation or hope of A to throw every time one six, at least, with six dice?

What is the expectation or hope of B to throw every time two sixes, at least, with twelve dice?

What is the expectation or hope of C to throw every time three sixes, or more than three, with eighteen dice?

And whether has not B and C as great an expectation or hope to hit every time what they throw for, as A hath to hit what he throws for?

If the question be thus stated, it appears, by an easy computation, that the expectation of A is greater than that of B or C; that is, the task of A is the easiest: and the reason is, because A has all the chances on sixes on his dice for his expectation, but B and C have not all the chances upon theirs; for, when B throws a single six, or C but one or two sixes, they miss of their expectations. This Mr. Smith understands, and therefore allows that, if the question be understood as I have stated it, then B and C have not so easy a task as A; but he seems of opinion, that the questions should be so stated, that B and C, as well as A, may have all the chances of sixes on their dice within their expectations. I do not see that the words of the question, as 'tis set down in your letter, will admit it; but this being no mathematical question, but a question what is the true mathematical question, it belongs not to me to determine it. I have contented myself, therefore, to set down how, in my opinion, the question, according to the most obvious and proper meaning of the words, is to be understood; and that, if this be the true state of the question, then B and C have not so easy a task as A: but, whether I have hit the true meaning of the question, I must submit to the better judgment of yourself and others. If you desire the computation, I will send it you.

I am, Sir,

Yr most humble and obedient Servant,

Is. Newton.

Isaac Newton to S. Pepys.

Cambridge, Decr. 16, 1693.

Sir—In stating the case of the wager, you seem to have exactly the same notion of it with me; and to the question, Which of the three chances should Peter chuse, were he to have but one throw for his life? I answer, that if I were Peter, I would chuse the first. To give you the computation upon which this answer is grounded, I would state the question thus:—

A hath six dice in a box with which he is to fling at least one six, for a wager laid with *R*.

B hath twelve dice in another box, with which he is to fling at least two sixes, for a wager laid with S.

C hath eighteen dice in another box, with which he is to fling at least three sixes, for a wager laid with *T*.

The stakes of *R*, *S*, and *T*, are equal; what ought *A*, *B*, and *C*, to stake, that the parties may play upon equal advantage?

To compute this, I set down the following progressions of numbers:—

Progr. 1.	1	2	3	4	5	6	the number of the dice.
Progr. 2.	0	1	3	6	10	15	
Progr. 3.	6	36	216	1296	7776	46656	{ the number of all the chances upon them.
Progr. 4.	5	25	125	625	3125	15625	{ the number of chances without sixes.
Progr. 5.	1	5	25	125	625	3125	
Progr. 6.	1	10	75	500	3125	18750	chances for one six and no more.
Progr. 7.		1	5	25	125	625	
Progr. 8.		1	15	150	1250	9375	{ chances for two sixes and no more.

The progressions in this table are thus found: the first progression, which expresses the number of the dice, is an arithmetical one; viz., 1, 2, 3, 4, 5, &c.; the second is found, by adding to every term, the term of the progression above it; viz., $0 + 1 = 1$, $1 + 2 = 3$, $3 + 3 = 6$, $6 + 4 = 10$, $10 + 5 = 15$, &c.; the third progression, which expresses the number of all the chances upon the dice, is found by multiplying the number 6 into itself continually; and the fourth, fifth, and seventh, are found by multiplying the number 5 into itself continually; the sixth is found by multiplying the terms of the first and fifth; viz., $1 \times 1 = 1$, $2 \times 5 = 10$, $3 \times 25 = 75$, $4 \times 125 = 500$, &c.; and the eighth is found by multiplying the terms of the second and seventh; viz., $1 \times 1 = 1$, $3 \times 5 = 15$, $6 \times 25 = 150$, $10 \times 125 = 1250$, &c.; and by these rules the progressions may be continued on to as many dice as you please.

Now, since *A* plays with six dice, to know what he and *R* ought to stake, I consult the numbers in the column under six, and there, from 46656, the number of all the chances upon those dice, expressed in the third progression, I subduct 15625, the number of all the chances without a six, expressed in the fourth; and the remainder, 31031, is the number of all the chances, with one six or above: therefore the stake of *A* must be the stake of *R*, upon equal advantage, as 31031 to 15625, or $\frac{31031}{15625}$ to 1; for their stakes must be as their expectations, that is, as the number of chances which make for them. In like manner, if you would know what B and S ought to stake upon twelve dice, produce the progressions to the column of twelve dice, and the sum of the numbers in the fourth and sixth progressions; viz., 244140625 + 585937500 = 830078125, will be the number of chances for S; and this number, subducted from the number of all the chances in the third progression, viz., 2176782336, will leave 1346704211, the number of chances for B: therefore the stake of B would be to the stake of S, as 1346704211 to 830078125, or $\frac{1346704211}{830078125}$ to 1. And so, by producing the progressions to the number of eighteen dice, and taking the sum of the numbers in the fourth, sixth, and eighth progressions for the number of the chances for T, and the difference between this number and that in the third column for the number of the chances for C, you will have the proportion of their stakes upon equal advantage. And thence it will appear that, when the stakes of R, S, and T, are units, suppose one pound or one guinea, and by consequence equal, the stake of A must be greater than that of B, and that of B greater than that of C; and, therefore, A has the greatest expectation. The question might have been thus stated, and answered in fewer words: if Peter is to have but one throw for a stake of 1000L., and has his choice of throwing either one six at least upon six dice, or two at least upon twelve, or three at least upon eighteen, which throw ought he to chuse; and of what value is his chance or expectation upon every throw, were he to sell it? Answer: Upon six dice there are 46656 chances, whereof 31031 are for him; upon twelve, there are 2176782336 chances, whereof 1346704211 are for him: therefore, his chance of expectation is worth the $\frac{31031}{46656}$th part of 1000L. in the first case, and the $\frac{1346704211}{2176782336}$th part of 1000L. in the second; that is, 665L. 0s. 2d. in the first case, and 618L. 13s. 4d. in the second. In the third case, the value will be found still less. This, I think, Sir, is what you desired me to give you an account of; and if there be any thing further, you may command

> Your most humble and most obedient Servant,
> Is. Newton.

Let us suppose that we have entered a lottery in which there are prizes of value £a, £b, £c, . . . , and let us also suppose that we know that the respective probabilities of obtaining these prizes by means of a single ticket are p, q, r, ... , respectively. If the lottery were drawn a large number N of times, the holder of a single ticket would win £a on pN occasions, £b on qN occasions, £c on rN occasions, ... Hence the holder of a single ticket in each of the N lotteries would receive

$$£(pNa + qNb + rNc + \ldots\ldots).$$

If he is to pay the same price £t for his ticket each time, we should have, for equity,

$$Nt = pNa + qNb + rNc + \ldots,$$

and therefore

$$t = pa + qb + rc + \ldots .$$

The price of his ticket is made up of parts corresponding to the various prizes, namely pa, qb, rc, ...

Calling these parts the *values of the expectations of the respective prizes*, we have the rule:

The value of the expectation of a sum of money is that sum multiplied by the chance of obtaining it.

We also see that *the value of the expectation of receiving any one of a number of prizes is the sum of the values of the expectations of receiving all of them.*

The first rule is never applied by those diligent people who, week after week, fill up football coupons. The prize they yearn for is of the order of £76,000, but the probability of winning it is one in many millions. Hence the value of their expectation is minute, and certainly much smaller than the sixpence which they expend on each entry. In other words, they systematically swindle themselves, week after week.

How much would one spend each week on a ticket which might ensure one's salvation? This may seem a crude notion, yet mathematical arguments have been seized on before now to prove theological beliefs. To quote Augustus De Morgan again:

"When a very young man, I was frequently exhorted to one or another view of religion by pastors and others who thought that a mathematical argument would be irresistible. And I have heard the following more than once. . . . . . :

"Since eternal happiness belonged to the particular views in question, a benefit infinitely great, then, even if the probability of their arguments were small, or even infinitely small, yet the product of the chance and the benefit according to the usual rule, might give a result which no one in prudence ought to pass over."

De Morgan adds that they did not see that this argument applied to all systems, as well as their own. He might have added that a cautious man would therefore subscribe to all possible theological doctrines which promised eternal happiness.

We now make use of this rule for finding the value of an expectation in another connection, equally controversial. As a rule, after checking the proof, mathematicians are disposed to accept results obtained by means of mathematics. But in the theory of probability there are some results which have given rise to much controversy, because of their paradoxical nature. We have had one paradoxical result already, but that was easily explained, and the mathematics which led to the result was incorrect. But the "St. Petersburg Paradox" is interesting because the result appears to be paradoxical, but the mathematics is correct.

This gambling problem was first proposed to Nicolaus Bernoulli in a letter dated September, 1713. It was modified by Daniel Bernoulli, nephew of Nicolaus, and discussed at length by him in the Transactions of the St. Petersburg Academy. Hence its name. The problem is the following:

A coin is tossed until heads appears. If heads appears on the first toss, the bank pays the player £1. If heads appears for the first time on the second toss, the bank pays the player £2. If heads appears for the first time on the third toss, the bank pays the player £4. If heads appears for the first time on the fourth toss, the bank pays the player £8, and so on. Now, what amount should the player pay the bank for the privilege of playing one game if the game is to be a fair one; that is, if neither player nor bank is to have an advantage, regardless of how long the game goes on?

To solve this problem we apply the rules obtained above. At the first toss the probability of heads is $\frac{1}{2}$, and the prize is £1. The value of the player's expectation is therefore $(\frac{1}{2}) (£1) = 10s$. The player will only win on the second toss if the first toss is tails, and the second heads. The probability of this is $(\frac{1}{2}) (\frac{1}{2}) = \frac{1}{4}$, and the prize is £2, so that the value of his expectation of winning on the second toss is $(\frac{1}{4}) (£2) = 10s$. The player will only win on the third toss if the first two tosses give tails, and the third heads. The probability of this is $(\frac{1}{2})^3 = \frac{1}{8}$. The sum to be won is £4, so that once again the value of his expectation of winning on the third toss is $10s$.

So we continue, and it is easy to see that the value of the player's expectation of winning on the nth toss is 10s., whatever the value of n. Hence, by the second rule, the value of the player's expectation of winning at the nth toss is

$$£(\tfrac{1}{2} + \tfrac{1}{2} + \ldots + \tfrac{1}{2}),$$

where there are n terms.

Now we recall that the game is to continue until heads turns up for the first time. Theoretically there is no limit to the number of tails which may appear before heads turns up, and this means that the above series of terms does not terminate. We do not need to know anything much about infinite series (Chapter VII) to see that the sum of a large number of halves is a large number, and that this number becomes larger than any number we care to name if the number of halves is large enough. In brief, the sum of the series is an infinite number, and *the player should therefore pay the bank an infinite sum for the privilege of playing the game*!

To which the reader will say "Nonsense!" Nobody in his right mind would think of paying any great amount of money for such an opportunity. Yet the mathematics is correct. Then what is wrong?

This question has been bothering mathematicians for some two hundred years. Condorcet and Poisson thought that the game contains a contradiction. The bank enters into an engagement which he cannot keep. If heads does not come down until the coin has been tossed 100 times, the player should receive £(2^{99}), which is an astronomically large sum, and the bank has deceived the player, for he would be unable to pay! Moral criticisms of the result have, in fact, been overabundant. Daniel Bernoulli, who first rescued the problem from oblivion, thought that the notion of mathematical expectation, with which the problem deals, should be replaced by that of *moral* expectation, in the calculation of which the worth of a fortune depends not only on its size, but on the satisfaction it can give!

Bernoulli suggested, in fact, that if a given fortune x is increased by an amount h, then the worth of the increase is h/x. For example, if you have only £10 and gain £1, your satisfaction is the same as that of a man who has £100 and gains £10, or that of a man who has £1,000 and gains £100. Working on this assumption, Bernoulli obtained a finite sum for the amount which the player should give the bank for equity. His theory brought Bernoulli as much fame as did his admirable works on physics, but it hardly explains the paradox.

Buffon, the eminent eighteenth-century naturalist, tested the result of the St. Petersburg problem by experiment. He played the game to a finish 2,048 times, and induced a child to toss the coin all through the experiment. These 2,048 games produced 10,057 crowns. There were 1,061 games which produced one crown each, 494 which produced two crowns, and so on. We shall give more details of Buffon's results later. In any case, they accorded with the reasonable expectation that a coin, when tossed, *would not continue* to come down tails.

After performing these experiments, Buffon objected to the theoretical result on several grounds. One objection raised the matter of the *time* it would take to play more than a certain number of games, a human lifetime possessing only a finite span! Another, and more relevant suggestion, was that any probability less than 1/10,000 should be considered to be zero. This suggestion, if adopted, would limit the game to a finite number of tosses, since the probability of tails turning up for n successive tosses is $(\frac{1}{2})^n$, and as soon as n is so large that

$$(\tfrac{1}{2})^n \text{ is less than } 1/10{,}000,$$

which involves the inequality:

$$2^n \text{ is greater than } 10{,}000,$$

the value of the player's expectation ceases to increase. If the bank is reasonable, and agrees that heads *must* turn up sometime, he might just as well pay up there and then! Since n is 14, the bank would pay £2^{13}, and the player would pay the bank £7 for the privilege of playing.

The explanations given so far do not accept the mathematics, which only contacts reality when it is understood that the game is supposed to be played *a large number of times*. In other words, if the player pays the bank a definite finite sum at the beginning of each game, and goes on playing, then, no matter how large this finite sum is, the player will *eventually* win very large sums of money. This was clearly seen by Bertrand, who said:

"If we play for pennies instead of pounds, for grains of sand instead of pennies, for molecules of hydrogen instead of grains of sand, the fear of becoming insolvent may be diminished without limit. This should not affect the theory, which does not require that the stakes should be paid before every throw of the coin. However much the bank may owe, the pen can write it. Let us keep the accounts on paper. The theory will triumph if the accounts confirm its prescriptions. Chance will probably, we can even say certainly, end by favouring the player. However much the player pays for the bank's promise, the game is to his advantage, and if the player persists, it will enrich him without limit. The bank, whether he is solvent or not, will owe the player a vast sum.

If we had a machine which could toss 100,000 coins a second, and register the results, and if the player paid £1,000 for each game, he would have to pay £100,000,000 every second. But in spite of all this, after several trillion centuries, he will make an enormous profit. The conditions of the game favour him, and the theory is right."

As this is a book on the art of the mathematician, we have thought it worth while to give Bertrand's courageous statement. It is evident that modern finance would have had no terrors for this Frenchman! One does want to know, however, how a coin can behave if it is repeatedly tossed. Buffon's experiment was repeated by several other scientists, and it was found that when a coin was tossed, until it came down heads, 8,192 times,

tails appeared   8 times running 17 times;
.... 9 .... 9 .... ;
.... 10 .... 2 .... ;
.... 11 .... 1 .... ;
.... 13 .... 1 .... ;
.... 15 .... 2 .....

These figures do induce the belief that anything which can probably happen will happen some time or other! The fact that the probability of an event is a very small number should never lead us to believe that it can never happen.

Buffon was a great naturalist. Between 1749 and 1788 there were published 36 volumes of his *Histoire Naturelle*. He also translated Newton's work on Fluxions, and experimented on burning mirrors. But he is assured of immortality by his discovery of a beautiful theorem in what is called *geometrical probability*.

We imagine a horizontal plane on which a set of equally-spaced parallel lines have been drawn (the seams between the planks on an ordinary wooden floor will do). Let the distance between the lines (the width of the planks) be a. A needle of length l is thrown down at random on the plane, and note is made of whether the needle meets one of the parallel lines. Buffon's theorem asserts that if the length of the needle is not greater than the distance between the lines, the probability of the needle meeting a line is $2l/a\pi$, where π is, of course, the ratio of the circumference of a circle to its diameter (see Chapter VII).

To simplify the experiment, we can choose a needle whose length is equal to the distance between the lines, when the probability becomes $2/\pi$. All that we have to do is to drop the needle at random N times, say, and to count the number of times the needle crosses a line. If this number be N', then the ratio N'/N, as N increases without limit, should tend to $2/\pi$. Hence a practical experiment, involving the simplest operations, should produce a value of π, which is one of the fundamental mathematical constants!

This is an experiment well worth doing, if the reader can think how to drop a needle "at random". It is doubtful whether a child is more random in its motions than an adult person, and although the needle need not be pointed to satisfy the theoretical requirements of the theorem, and therefore a legitimate fear that the child may injure itself need not enter into the question, the disadvantages of employing child labour probably outweigh the advantages, and the precedent of George Louis Leclerc, Comte de Buffon, need not be followed in testing Buffon's Needle Theorem.

In 1901 the Italian mathematician Lazzerini dropped a needle 3,407 times, and obtained a value of π equal to 3.1415929, which is in error by less than 0.0000003. It is not known to the author how Lazzerini ensured a random dropping. Perhaps he too whirled around and then suddenly let go. If so, he must have become rather giddy before he had finished! There are many reasons for thinking that Lazzerini's accuracy is too good to be acceptable.

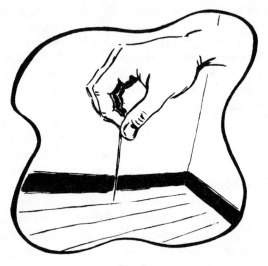

Fig. 5.

So far the author has probably not offended too many people in his discourse on probability. But now he is about to venture into very troubled waters. In Duke University, North Carolina, experiments have been carried out by Professor Rhine and his colleagues on extra-sensory perception and psycho-kinesis. The theory which has developed rests entirely on the interpretation of dice-throwing and card-shuffling experiments.

For example, a single die is thrown, and someone tries to *influence* the die, without actually handling it, so that it falls with the six uppermost. Now, *on the average,* the die should show a six once in six times. This is the only point at which probability enters, and we have said enough to show that such a statement is a very complex one. In the Rhine experiments sequences, or runs of throws are made, say one hundred at a time, and the ratio of the number of sixes to the total number of throws is noted.

It is found in some cases that this ratio is consistently greater than $\frac{1}{6}$ for a number of runs, but that it then drops off, and becomes consistently less than $\frac{1}{6}$. The fact that the ratio is greater than $\frac{1}{6}$ in some cases is proof, to the North Carolina school, that the die can be influenced by a person at a distance. Hence the theory of psychokinesis. The dropping-off of the scoring rate worried the experimenters for years, until someone had a brilliant notion. Since the scoring-rate should be neither greater nor less than $\frac{1}{6}$, they argued, the persistence of a score *less* than $\frac{1}{6}$ was also indubitable proof of the triumph of mind over matter! But why anyone wishing hard for sixes to turn up should, after a time, go into reverse gear, as it were, this has not been explained.

In every account of the Rhine experiments we see the phrase: "The mathematics is certainly correct." But our account of the St. Petersburg paradox shows that mathematics has to be *interpreted*, especially in probability theory. All that mathematics can say about the throwing of a single die is that *in the long run* a six will turn up for $\frac{1}{6}$ of the total number of throws, provided that the throws are random throws, and by this time the reader may well wonder whether such a statement has any significance at all. There is much in probability theory which we have not been able to discuss. It is a living subject, of the greatest theoretical and practical importance, and the more thought given to it, the better.

In our next chapter we shall see how mathematicians handle *infinity*.

CHAPTER III

WHERE DOES IT END?

THE infinite appears in a variety of forms in mathematics. It has already appeared in this book. Mathematicians, as we have seen, are not afraid of thinking of very large numbers, but any *definite* number, however large, is considered *finite*, not infinite. Thus the population of the world at any instant is a finite number, and so is the number of grains of sand on all the beaches of Britain. On the other hand, the number of natural numbers, 1, 2, 3, 4, ... and so on, is infinite because, however far one goes in counting, say up to a number N, the number $N + 1$ is another natural number which has not yet been allowed for. As we shall see later, the class of natural numbers, or integers, is one of the fundamental measures for infinity.

Another example of an infinite class, or set, of numbers is given by the *squares* of all the natural numbers:

$$1, 4, 9, 16, 25, 36, 49, \ldots \ldots$$

That this is an infinite class is proved in the same way. If we hopefully arrive at the number n^2, we cannot stop there, because there is also a number $(n + 1)^2$ in the class. Hence we have two classes, each containing an infinity of natural numbers, and it is natural to enquire whether one of the classes contains "more" numbers than the other.

It is certainly true that all members of the second class are *contained* in the first, since n^2, which is a typical member of the second class, is also an integer, and therefore a member of the first class. Again, there are many members of the first class which are not perfect squares, and are therefore not members of the second class. Can we not therefore say, that in spite of the fact that both classes contain an infinite number of members, the first class somehow contains a *greater infinity* than the second?

This very problem was discussed by Galileo in his *Dialogues*. He came to the conclusion that all that we can say about the two classes is that each of them is infinite, and that the relations of equality and inequality can be applied to finite, but not to infinite classes. There the matter rested until the possibility of comparing *degrees of infinity* was realised by Cantor, a German mathematician born in Russia, in

1873. Out of his work a most astonishing branch of mathematics has developed. The fundamental ideas are extremely simple.

In order to understand Cantor's reasoning, we must begin with *counting*. What do we mean when we say that there are twenty-one members in a finite set of objects? It is not enough to answer that we point at each member of the class in turn, and recite "One, two, three, ... , twenty, twenty-one." The ability to perform this operation indicates the possession of a highly specialised vocabulary of number words. How could we convey the sense of twenty-one

Fig. 6.

objects to someone who does not understand our language, and whose own language is so undeveloped that there are no words for "five", "six", ... , the only number words being "one", "two", "three" and "four", anything above "four" being many?

We could obviously convey the sense of "twenty-one" by cutting this number of notches on a stick, or by depositing this number of beads on the ground, or by holding up both hands with all the fingers spread out twice in succession, and then holding up one solitary finger. All these methods will be familiar to the reader from books he has read, or from his own experience. There is no doubt that the *sense* of the number twenty-one can be conveyed.

What we have done, in technical language, is to set up a "one-to-one" correspondence between the marks cut on a stick, or the beads, or our fingers and the objects whose *number* is twenty-one. This is counting in its simplest and most fundamental form. For every object there is *one* notch on the stick, *and* every notch on the stick corresponds to just *one* object.

We consider another example of one-to-one correspondence. I know that the room I teach in contains twenty-one desks. If, when I enter the room, I see that all the desks are occupied, then I know that twenty-one students have decided to attend that particular class. If, on the other hand, some of the desks are empty, then fewer than twenty-one students are present. If, finally, all the desks were occupied, and some students were standing, I should know that more than twenty-one students had turned up, and I should need more desks.

Hence a one-to-one correspondence between desks and students indicates the *same* number of desks and students. If there are seats to which no students belong, there is a *larger* number of desks than students. But if all the desks are occupied, and some students have no desks assigned to them, then the number of students is *greater* than the number of desks.

This is all very obvious, but it contains the germ of Cantor's great idea. The *actual number* of desks does not enter into the notion of equality or inequality. Hence we say:

If two infinite classes are such that a one-to-one correspondence can be set up between their members, then the two classes have the *same transfinite number* of members. This defines equality.

If the sets be M and N, we write $M \sim N$ to symbolize the fact that M and N can be put in one-to-one correspondence.

If U be another set, and $N \sim U$, then it is clear that $M \sim U$. Hence from $M \sim N$ and $N \sim U$, we deduce that $M \sim U$. The symbol behaves, so far, like the $=$ sign.

When can we say that the transfinite number, α, say, of a set M is *less* than the transfinite number β, say, of a set N? Two conditions must be satisfied:

(1) There exists no proper subset M_1 of M such that $M_1 \sim N$, and

(2) There exists a proper subset N_1 of N such that $N_1 \sim M$.

A *proper* subset of a set M is a set contained in M which does not coincide with it. Condition (2) is a natural condition which is illustrated in the finite case by the example already given of desks and students. Condition (1) is essential because we are dealing with infinite sets, as the following examples show. But we pause for a moment longer to show that conditions (1) and (2) are inconsistent with $\alpha = \beta$.

For if $\alpha = \beta$, then $M \sim N$. But since $N_1 \sim M$, it follows that $N_1 \sim N$.

Returning to $M \sim N$, we deduce that there exists a proper subset M_1, say, of M which is such that $M_1 \sim N$, and this contradicts condition (1).

With these new concepts in mind, we return to the infinite class which consists of the natural numbers

$$1, 2, 3, 4, 5, 6, 7, \ldots \ldots$$

We can set up a one-to-one correspondence between the natural numbers and their squares

$$1, 4, 9, 16, 25, 36, 49, \ldots.$$

by making any number n in the first class correspond to n^2 in the second class, and making any number n^2 in the second class correspond to n in the first class, thus:

$$1 \quad 2 \quad 3 \quad 4 \quad 5 \quad 6 \quad 7 \ldots \ldots n \ldots \ldots$$
$$1 \quad 4 \quad 9 \quad 16 \quad 25 \quad 36 \quad 49 \ldots \ldots n^2 \ldots \ldots$$

It is true that we cannot demonstrate this correspondence for *every* member of each class, but it is sufficient to demonstrate it for a *typical* member n of the first class, and a typical member n^2 of the second class.

Having demonstrated the existence of a one-to-one correspondence, we conclude that the class of the squares of all the natural numbers has the same transfinite number as the class of all the natural numbers! This result is not what might have been anticipated, seeing that the second class is a proper subset of the first.

Similarly, the class of all *even* numbers has the same transfinite number as the class of all natural numbers. The one-to-one correspondence looks like this:

$$1 \quad 2 \quad 3 \quad 4 \quad 5 \quad 6 \quad 7 \ldots \ldots n \ldots \ldots$$
$$2 \quad 4 \quad 6 \quad 8 \quad 10 \quad 12 \quad 14 \ldots \ldots 2n \ldots \ldots$$

Again, the class of all *odd* numbers has the same transfinite number as the class of natural numbers, the one-to-one correspondence being

$$1 \quad 2 \quad 3 \quad 4 \quad 5 \quad 6 \quad 7 \ldots \ldots n \ldots \ldots$$
$$1 \quad 3 \quad 5 \quad 7 \quad 9 \quad 11 \quad 13 \ldots \ldots 2n-1 \ldots.$$

In each of these three examples the class of natural numbers has been put in one-to-one correspondence with a *part of itself*. In other words, we have been demonstrating that *the whole is equal to part of itself*. This is in direct contradiction to the familiar assumption, or axiom, first encountered in geometry, that *the whole is equal to the sum of its parts, and is therefore greater than any of them*. This axiom, of course, refers to *finite* magnitudes.

That the whole is equal to part of itself, paradoxical as it may seem, is a conclusion which involves the essence of infinite magnitude. We have not, the reader may have noticed, defined an *infinite class* as yet. But we can now, with Cantor, *define* an infinite class as a class which can be put into one-to-one correspondence with a part (or proper subset) of itself!

If the reader feels that this definition is unnecessarily abstract, and that it is always possible, by *counting* the members, to see whether a class is infinite or not, he must be warned that we shall soon produce a class whose members *cannot be counted*! But before we produce this specimen, we show that the class of all positive rational numbers p/q, where p and q are integers, *can* be counted. This means that the class has the same transfinite number as that of the natural numbers.

It is high time that we gave this last transfinite number a name, and we cannot do better than to follow Cantor, and to use the first letter *aleph* $\aleph$ of the Hebrew alphabet (Cantor actually used $\aleph_0$) to denote the transfinite number of the class of natural numbers. Classes which have the transfinite number $\aleph$ are said to be *countable*, or *denumerable*.

It is surprising that the class of all positive rational numbers turns out to be denumerable, because we can interpose an infinity of rational numbers between any two given rational numbers. For instance, between 0 and 1 we can interpose the rationals

$$\tfrac{1}{2}, \tfrac{2}{3}, \tfrac{3}{4}, \tfrac{4}{5}, \ldots \ldots, n/(n + 1), \ldots;$$

between 0 and $\tfrac{1}{2}$ we can interpose

$$\tfrac{1}{3}, \tfrac{2}{5}, \tfrac{3}{7}, \tfrac{4}{9}, \ldots \ldots, n/(2n + 1), \ldots;$$

and so on. Because of this property we might well expect that the transfinite number of the class of all positive rationals would exceed $\aleph$. But Cantor showed that the positive rationals can be counted, and we now give his proof.

We arrange the positive rationals as shown below. In the first row all the numerators are 1, and the successive denominators are 1, 2, 3, 4, ... In the second row all the numerators are 2, and the successive denominators are 1, 2, 3, 4, ... , as before. So we continue. In the nth row the numerators are all n, and the successive denominators are 1, 2, 3, 4, . . .

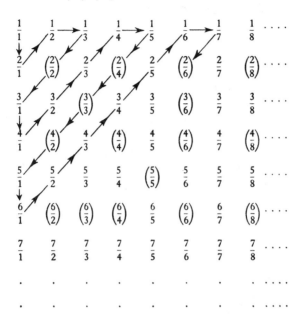

We have enclosed in brackets all the fractions which have a factor common to numerator and denominator. If these particular fractions be deleted, then every rational number appears once, and once only in the above array of positive rationals. All that we need to do now is to indicate how the numbers can be counted, our method being such that no rational number is omitted from the process.

The arrows in the diagram indicate the order of counting, that is, the way in which we set up a one-to-one correspondence between the natural numbers and the positive rationals. These are paired as shown below

$$1 \quad 2 \quad 3 \quad 4 \quad 5 \quad 6 \quad 7 \quad 8 \quad 9 \quad 10 \quad 11 \quad 12 \quad 13 \ldots\ldots$$

$$\frac{1}{1} \quad \frac{2}{1} \quad \frac{1}{2} \quad \frac{1}{3} \quad \frac{3}{1} \quad \frac{4}{1} \quad \frac{3}{2} \quad \frac{2}{3} \quad \frac{1}{4} \quad \frac{1}{5} \quad \frac{5}{1} \quad \frac{6}{1} \quad \frac{5}{2} \ldots\ldots$$

Of course, the one-to-one correspondence in this case is not as simple as in the cases already considered, and in fact it would not be easy to obtain a formula for the *position*, in this enumeration, of the fraction

p/q. But the particular way in which a one-to-one correspondence is set up is immaterial. The important thing is to exhibit some sort of systematic method for pairing the positive rationals with the natural numbers, and this we have done. To recapitulate, we first arrayed the rational numbers in a scheme in which every positive rational number appears once, and only once. We then indicated how to pair every rational number with an integer. If the rational number is named we can, by going along the diagonal path indicated by the arrows, find the integer which corresponds to it. On the other hand, if the integer be given, we can count as we proceed along the path until we come to the unique rational number which is assigned to it. We have therefore shown that the positive rational numbers are denumerable.

The above proof is one of the great proofs of modern mathematics. Although it is, at first glance, a visual proof, it is completely rigorous. It also appears to be as inevitable as any other great work of art. The more it is contemplated, the more one's admiration grows. The next theorem, which is also due to Cantor, is perhaps even more wonderful, but is not the kind which commands universal approbation. It is necessarily indirect, and, as we shall see in Chapter VIII, there are mathematicians who will not accept proofs by *reductio ad absurdum*. But before we criticise, let us see what the theorem is.

We prove that there is an infinite class which has a transfinite number greater than $\aleph$, and is therefore not denumerable. This class is the class of all *real numbers* between 0 and 1. From our point of view a real number is one which has a decimal expansion, so that the class of real numbers between 0 and 1 are all those which are of the form

$$0 . a_1 a_2 a_3 \ldots \ldots \ldots ,$$

where the sequence of digits after the decimal point may terminate or be infinite.

To prove that the class of real numbers lying between 0 and 1 is not denumerable we assume the contrary, and show that this leads to a contradiction. We assume, then, that it is possible to establish a one-to-one correspondence between the real numbers between 0 and 1 and the natural integers. We shall then show that there is a real number which has not been counted, and has no place in the enumeration.

Before we begin the proof we prepare our real numbers. As we shall see in Chapter VII, a number has a *unique* decimal expansion unless the expansion terminates after a finite number of decimal places, when it may also be represented by an infinite decimal expansion, involving the sequence .999999. . . . . For example, the number .345 may also be represented by the decimal .3449999. . . . . This is because 0.999 . . . . . = 1. If therefore a number has an infinite decimal expansion, no other infinite decimal expansion can represent the same number. A difference in any one digit in two infinite decimal expansions means that the expansions represent two distinct numbers. We wish to use this fundamental property of decimal expansions, and therefore we see to it that all numbers to be considered are clad in an infinite decimal expansion.

The assumption is that all decimals of the form . $a_1 a_2 a_3$ . . . . . can be put in order. Let this order be as shown below:

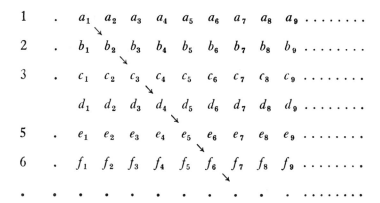

We assume, of course, that the numbers on the right are all known, and that we have a list which extends as far as we wish it to extend. We now show that there is a decimal lying between 0 and 1 which *does not appear anywhere in the list*. This decimal is *constructed* as follows; we shall call it . $z_1 z_2 z_3 z_4$ . .

If a_1 is any one of the digits 0, 1, 2, 3, 4, 5, 6, 7, then z_1 shall be 8. If a_1 is either 8 or 9, then z_1 shall be 1.

If b_2 is any one of the digits 0, 1, ... , 7, then z_2 shall be 8. If b_2 is either 8 or 9, then z_2 shall be 1.

If c_3 is any one of the digits 0, 1, 2, . . , 7, then z_3 shall be 8. If c_3 is either 8 or 9, then z_3 shall be 1.

The procedure should now be quite clear. This decimal we have written down lies between 0 and 1, but where is its place in our enumeration? It cannot be the first decimal, since, by construction z_1 differs from a_1. It cannot be the second decimal, since by construction z_2 differs from b_2. It cannot be the third decimal, since, by construction z_3 differs from c_3, ... It cannot be the nth decimal since, by construction, z_n differs from the nth digit of the decimal in the nth place in the enumeration.

Hence the decimal we have constructed has no place in the enumeration, and the claim that all decimals lying between 0 and 1 have been put in order must be false. This proves that there is an infinite class of numbers whose members, in the ordinary sense, cannot be counted. This class contains, as a proper subset, the class of rationals

$$\frac{1}{1} \quad \frac{1}{2} \quad \frac{1}{3} \quad \frac{1}{4} \quad \frac{1}{5} \quad \cdots \cdots \quad \frac{1}{n} \quad \cdots \cdots$$

which evidently has the transfinite number $\aleph$. We are therefore justified in saying that the transfinite number of the class of real numbers lying between 0 and 1 is *greater* than $\aleph$.

Is there an infinite class whose transfinite number lies *between* $\aleph$ and the transfinite number of the class of real numbers lying between 0 and 1? It is *believed* that there is not, but this has not yet been proved, and must be regarded as one of the great unsolved problems of modern mathematics.

Perhaps enough has been said to show that Cantor was one of the *enfants terribles* of mathematics. Mathematicians are still divided into two classes, those who swear by him, and those who swear at him! In the 1930's there were to be found reputable mathematicians in Germany who could (so they asserted) detect a non-Aryan mathematician by the way he approached certain branches of mathematics. These pure Aryans found modern mathematics decadent, of course, and tried to convince the world that the fault lay with Cantor. But on the other hand, the greatest mathematician of that epoch, Hilbert, was unrestrained in his praise of Cantor, and spoke with feeling of the paradisial delights which Cantor had created for the mathematicians of all time. More will be said about those who object to these paradisial delights, on what may be called mathematically ethical grounds, in Chapter VIII.

We shall consider the problems of infinite series, including infinite decimals, in Chapter VII. Their consideration involves a fair amount of mathematical technique. In this chapter we have shown that some fundamental notions of infinity are simple enough for everyone to understand. In our next chapter we switch to a discussion on automatic thinking, in which symbols do the work for us, if we control them judiciously.

CHAPTER IV

AUTOMATIC THINKING

A FAIR amount of mathematical thinking is mechanical, or automatic. If a problem can be stated in terms of symbols, and if the law of operation of these symbols is known, a process can be set in motion which will eventually lead to the desired solution. The reader must surely remember the rules he was taught in elementary algebra: "Change the sign when you take over to the other side; minus multiplied by minus gives plus;", and so on. We shall be considering these rules in Chapter VI, but here we deal with a different set, which are simpler, more fundamental perhaps, and, when applied, solve some amusing and interesting problems which are more difficult to solve by other methods.

We have already discussed *classes* of numbers, in the preceding chapter. By a *class* of objects or individuals we shall mean those objects or individuals which possess a certain property. For example, the property might be that of having only one leg, and refer to men, in which case the class would consist of all men with only one leg. This class does not, fortunately, contain all men. If we think of the class of all men, the class of men with only one leg lies in it, since a man with one leg is also a man. If we call the class of all men with one leg x, and the class of all men y, we say that the class x *lies in*, or is *contained in* the class y because every member of x is a member of y. In symbols we write this

$$x \subset y,$$

or
$$y \supset x.$$

Now the class of all men is contained in the class of all human beings, which we may denote by z. Hence we may write

$$y \subset z.$$

It is clear that a man with a single leg is a human being, and therefore

$$x \subset z.$$

Hence the relations $x \subset y$, and $y \subset z$ lead to the relation $x \subset z$.

This automatic deduction of a third relationship from two given ones is perhaps as simple as any deduction can be, but it will help us to solve some complicated problems. To make the matter as intuitive as possible, let us use diagrams as well. We represent the members of a class x, say, by the points in and on the circumference of a circle. The members of a class y are represented by the points in and on the circumference of another circle. If $x \subset y$, then the circle representing the class x lies inside the circle representing the class y.

If $y \subset z$, the circle representing the class y lies inside the circle representing the class z, and it is evident that the circle representing the class x lies inside the circle representing the class z.

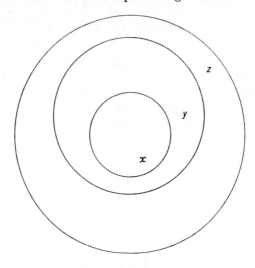

Fig. 7.

Most people have, at some time or another, come across the following statement: "All men are mortal; Socrates was a man; therefore Socrates was mortal." It is only nowadays, when logic is no longer taught to schoolboys, that it cannot be assumed that every schoolboy knows that the above sequence of sentences constitutes an Aristotelian *syllogism*. The study of the various types of syllogism constituted an essential part of logic. Without any symbolism to help the student, such a training was a severe one. Here we show how our simple mathematical symbolism works in the case of this syllogism and some complicated modern variants.

Let M stand for the class of mortals, and let m stand for the class of men. Since all men are mortal, the class of men is contained in the class of mortals. We may therefore write

If S denotes Socrates, then $S \subset m$, since Socrates was a man. We therefore have the relationships

$$S \subset m \subset M,$$

from which we deduce that $S \subset M$; that is: Socrates was mortal.

Before we consider the next example, which is not quite so obvious, we introduce some new concepts. We denote the class of those individuals which are *not* members of the class x by the symbol x'. This can be read as "not x". If the class x is represented by the points within and on the circumference of a circle, the class x' is represented by the points outside the circle. A diagram shows that if $x \subset y$, then $x' \supset y'$.

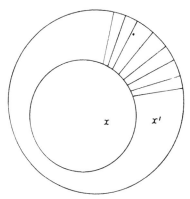

Fig. 8.

The reader may well wonder how far the region outside the circle which represents the class x, say, is to extend. In any particular problem there is a class which consists of *all the individuals under discussion*. We call this the *universal class*, and can represent this by a circle in the diagram. Then, in the above case, the circles representing the classes x and y both lie inside this circle, and the class x', or "not x" is represented by the points inside the largest circle and outside the circle representing x. It is then even clearer that $x' \supset y'$ follows from $x \subset y$, and conversely.

As a final definition at this stage, we define the *null-class* as the class with no members, or the class whose members do not exist within the universal class. If the universal class consists of all normal human beings, there are no members of the class which consists of all men with three legs, and such a class is *empty*. We shall denote the null-class by 0, which is not a number in this case, but a symbol. There will be no confusion since, for the moment, no operations will be carried out on this symbol. It will merely represent the null class. Now for a slightly more complicated example:

"Some laws are complicated; no confusing law is satisfactory; every complicated law is confusing." What deduction do we make from this set of given facts, or premises?

The universal class here is the class of all laws. Let b denote the class of all complicated laws, c the class of all confusing laws, and d the class of all satisfactory laws. We are told that the class b is not the null set, since some laws are complicated. We are also told that confusing laws are not satisfactory. Since d represents the class of satisfactory laws, d' must represent the class of *un*satisfactory laws, and

$$c \subset d'.$$

Finally the last assertion gives us

$$b \subset c.$$

Hence we have

$$b \subset c \subset d',$$

from which we deduce that

$$b \subset d',$$

which means that the class of complicated laws is contained in the class of unsatisfactory laws. Since the class of complicated laws is not empty, we can say:

Some laws are unsatisfactory.

The reader may still feel that he can do without this symbolism, but naturally we begin with easy examples. Let us consider the following, from the Reverend Charles Lutwidge Dodgson's "Symbolic Logic":

(1) All writers who understand human nature are clever:
(2) No one is a true poet unless he can stir the hearts of men:
(3) Shakespeare wrote Hamlet:
(4) No writer who does not understand human nature can stir the hearts of men:
(5) None but a true poet could have written Hamlet.

This is a compound syllogism. To make a legitimate deduction from the given premises we note that the universal class here consists of "writers".

We denote by a the class able to stir the hearts of men: by b the class of clever writers: c stands for Shakespeare: d for the class of writers who are true poets: e for the class of writers who understand human nature, and finally h for the writer of Hamlet.

We now interpret the statements symbolically; they become

(1) $e \subset b$: (2) $d \subset a$: (3) $c = h$: (4) $e' \subset a'$: (5) $h \subset d$.

Only the symbolism for (4) requires some explanation. Writers who do *not* understand human nature are represented by e', and since they cannot stir the hearts of men, they are contained in the class a', of writers unable to stir the hearts of men. Another way of interpreting (4) is to write $a \subset e$.

We can now write down the sequence

$$c = h \subset d \subset a \subset e \subset b.$$

This makes use of all the relations, and leads to the conclusion

$$c = h \subset b,$$

or

Shakespeare was a clever writer.

As a final example of the use which can be made of the inclusion-relation, we solve the following ten-part syllogism, which is even more in the Lewis Carroll vein:

(1) The only animals in this house are cats;
(2) Every animal is suitable for a pet, that loves to gaze at the moon;
(3) When I detest an animal, I avoid it;
(4) No animals are carnivorous, unless they prowl at night;
(5) No cat fails to kill mice;
(6) No animals ever take to me, except what are in this house;
(7) Kangaroos are not suitable for pets;
(8) None but carnivora kill mice;
(9) I detest animals that do not take to me;
(10) Animals, that prowl at night, always love to gaze at the moon.

The universal class here consists of "animals". All the following letters stand for classes of animals:

a = avoided by me; b = carnivora; c = cats; d = detested by me; e = in this house; h = kangaroos; k = killing mice; l = loving to gaze at the moon; m = prowling at night; n = suitable for pets; r = taking to me.

We write down the expression of the given premisses:

(1) $e \subset c$: (2) $l \subset n$: (3) $d \subset a$: (4) $b \subset m$: (5) $c \subset k$:
(6) $r \subset e$: (7) $h \subset n'$: (8) $k \subset b$: (9) $r' \subset d$: (10) $m \subset l$.

The statement of the given premises in symbolical form is simple when it is realised that (4), for example, is equivalent to: animals which are carnivorous prowl at night, and that (5) means: all cats kill mice, and so on.

Naturally the premises are not stated in the form which we, for our purpose here, should regard as the simplest, and most of the work in examining premises consists in releasing the underlying structure from the verbal creepers which entangle it. Now, as soon as we have sorted out the ten symbolical statements given above, we see that

$$a' \subset d' \subset r \subset e \subset c \subset k \subset b \subset m \subset l \subset n \subset h',$$

and taking the first and last, we deduce that

$$a' \subset h',$$

which is equivalent to

$$h \subset a,$$

which reads:

Fig. 9.

kangaroos are included in the class of animals avoided by me, or, more simply:

I avoid kangaroos!

The work done in solving this example was purely mechanical, and exactly equivalent to putting a number of objects in line in ascending order of size. Of course the example "comes out", so that we are able to make a deduction from the given premises, and we are not led to a contradiction. More subtle examples need a more subtle symbolism, and we shall develop an *algebra of classes*. This was first done by George Boole, nearly a hundred years ago, and described in his book *An Introduction to the Laws of Thought*. The algebra of classes is called *Boolean algebra*, in his honour. Our reason for developing it is that we shall be able to use it to solve problems, very fashionable these days, which do not yield to ordinary algebra, which we discuss in Chapter VI.

We have already considered classes of objects. Let x, say, denote the class of men with one leg, and let y denote the class of men with brown eyes. How shall we denote the class of men with one leg *and* with brown eyes?

We find that a suitable notation for such a class is the *product xy*. The same class can also be denoted by yx, the class of men with brown eyes and one leg, so that in this algebra we have

$$xy = yx.$$

This is a comfort! Now, if we wish to consider the class which has property x *or* property y, or both, we find that we can designate it by $x + y$. Thus the class of men with either one leg, or brown eyes, or both, in our example, is represented by $x + y$. Here again we have

$$x + y = y + x.$$

If z denotes a class of individuals with a common third property, the symbols $(xy)z$ and $x(yz)$ are easily interpreted. Each stands for the class of those individuals possessing all three properties, that of the class x, and the class y, and the class z, so that

$$(xy)z = x(yz),$$

and we may write this class, without ambiguity, as xyz.

Again

$$(x + y) + z$$

stands for the class which has the property x, or the property y, or the property z, the "or" not being exclusive, so that individuals which have both properties x and y, or y and z, or z and x, or x and y and z are included. Hence

$$(x + y) + z = x + (y + z),$$

and we can write $x + y + z$ for this class, without ambiguity. Our next step is the fundamental one which relates multiplication and addition. Of course, these are not the ordinary operations of arithmetic, but we are guided by the pattern which we know already exists in arithmetic.

In the algebra of classes we have the rule:

$$x(y + z) = xy + xz.$$

This is called the *distributive* law, as in algebra, and must be *proved*, since we have *defined* multiplication and addition, and this law connects the two operations. The proof consists in interpreting both sides of the equation. The left-hand side consists of those individuals which have property x *and* property $y + z$; that is property x and property y, or property x and property z, the "or" not being exclusive. But the right-hand side defines precisely the same class of individuals. Hence the two classes are the same, which is what the equation means.

This is the only rule which we shall use in what follows. Some special consequences, which follow because of the special nature of the objects represented by our symbols, must, however, be indicated. What do we mean by a product $x \cdot x$, which we write as x^2? This is the symbol which represents a class which has the properties of the class x, and of the class x! But this class cannot be distinguished from the class x, since saying a thing twice, if we are not trying to invoke the spirits of the air, does not make it more potent. Hence

$$x^2 = x.$$

In the same way

$$x^3 = x,$$

and so on. Similarly

$$x + x = 2x = x,$$
$$x + x + x = 3x = x,$$

and so on.

The null class, which is empty, has already been denoted by 0. We attach a symbol to the universal class also, the class which consists of all objects under discussion. We shall denote the universal class by the symbol 1.

It has already been stressed that the 0 and 1 we are using here are not the 0 and 1 of ordinary arithmetic, but we find that *they do obey similar rules*. In fact we have

$$0 \cdot x = 0, \qquad 1 \cdot x = x,$$
$$0 + x = x.$$

The first equation merely states that there are no individuals common to the null-class and the class x, so that the product is the null-class. The second rule follows because the class x is contained in the universal class, and so the elements common to both classes are precisely those of the class x. The third rule states that the class which is defined to have either the properties of the class x, or the properties of the class which has no members, is merely the class x.

We defined x' as the class which does not have the properties of the class x. We have the following consequences:

$$x \cdot x' = 0: \quad x + x' = 1: \quad (x')' = x:$$
$$1' = 0: \quad 0' = 1.$$

The interpretation of these is not difficult. The first result merely states that there are no individuals which satisfy a property *and* its negation at the same time. The second result says that the individuals which satisfy a given property, together with those which do not, make up all the individuals we are considering.

The third result states that a double negative is equivalent to a positive: that is, the individuals *not* satisfying the property of *not* satisfying a given property are precisely those which satisfy the given property. This rule is sometimes referred to as "the law of the excluded middle", and is based on the assumption that a statement is either true or false. We shall come across it again in Chapter VIII.

The last two results are evident. The diagram introduced above helps to clarify these algebraic results. As we said, the points *outside* the circle which represents the class x represent the class x'. Hence the first result states that there are no points which are both outside and inside a given circle, and so on.

The reader will probably feel more at ease with *propositions* than with classes. From our point of view a proposition is a sentence which conveys a meaning which is true or untrue. For example, the proposition may be "Oxford Street is in London", or "Bankers have wings". We are going to deal with the algebra of propositions, but there is no need for alarm! It is exactly similar to the algebra of classes developed above. If we use the symbol p for our first proposition, and q for the second, then the product pq will denote *the two propositions taken in conjunction*; that is it denotes a proposition which asserts both p and q *simultaneously*. In the example we have given, pq would be false since, although p is true, the proposition q would be denied by most people.

The symbol $p + q$ is taken to denote the propositions p and q taken *in disjunction*; that is, it denotes the proposition asserting *either p or q*, the "either" not being exclusive. In the example given, $p + q$ is a true proposition, since p is true.

In the algebra of classes it is evident what we mean when we say that a class $x = $ a class y. In the algebra of propositions we say that the proposition $p = $ the proposition q when the two propositions are *logically* equivalent. We can now prove the distributive law

$$p(q + r) = pq + pr.$$

The left-hand side asserts the proposition p and the proposition $q + r$; that is the proposition p and the proposition q or the proposition r; which is the same as the proposition p and the proposition q or the proposition p and the proposition r; which is what the right-hand side asserts.

If p denotes a proposition, then p' will denote the *negation* of p, or "not- p". We introduce one new idea at this stage. We say that a true proposition has *the truth-value* 1, and if we write

$$p = 1,$$

this will mean "p is true". We also say that a false proposition has the truth-value 0, and if we write

$$p = 0,$$

this will mean "p is false". All the remaining results of the algebra of classes now hold in this algebra, and are easily interpreted. For example

$$pp' = 0, \qquad\qquad p + p' = 1$$

mean respectively:

> a statement p cannot be true and false at the same time:

> a statement p is either true or false.

If the statement p has the truth-value 1, and the statement q the truth-value 0, the two statements or propositions p and q taken in conjunction have the truth-value 0, as we should expect from $0 . 1 = 0$. But in this algebra if $pq = 1$ we can deduce that both p *and* q have the truth-value 1, since if either p or q were false, the product of their truth-values would be 0.

It is high time that we saw what kind of problem can be solved by the use of this algebra. We give an example, due to Hubert Phillips:

Alice, Brenda, Cissie and Doreen competed for a scholarship. "What luck have you had?" someone asked them. Said Alice: "Cissie was top, Brenda was second." Said Brenda: "No, Cissie was second, and Doreen was third." Said Cissie: "Doreen was bottom, Alice was second."

Each of the three girls made two assertions, of which only one was true. Who won the scholarship?

Now trial and error, or the systematic enumeration of cases, will solve puzzles of this kind. Ordinary algebra cannot touch them. But the algebra we have discussed above enables such problems to be approached with a standard technique, and has the advantage that it will reveal any alternative answers which a solution by trial and error tends to overlook. Finally, once the machine is set in motion it automatically churns out the answer.

In this problem each of three girls makes two statements, and one of the two statements is known to be false. How can we symbolise this? If p and q represent propositions, and one is known to be false, but only one, we have

$$p + q = 1 \quad \text{and} \quad pq = 0.$$

Neither of these is sufficient by itself, since the first equation is satisfied if *both* p and q are true, and the second is satisfied if *both* p and q are false. We can combine these two equations into the one equation

$$pq' + p'q = 1,$$

which merely states that:

either p and the negation of q are true, or q and the negation of p are true.

This effectively sums up our information. Each statement by the three girls is of this kind, and all we have to do now is to symbolise these statements.

We let A_1 stand for the proposition "Alice was first", A_2 for the proposition "Alice was second", and use initial letters with suffixes to denote any other propositions we may require. For example, D_4 stands for the proposition "Doreen was fourth". We now have the following information:

From Alice:

$$C_1 B'_2 + C'_1 B_2 = 1;$$

From Brenda:

$$C_2 D'_3 + C'_2 D_3 = 1;$$

From Cissie:

$$D_4 A'_2 + D'_4 A_2 = 1.$$

Since the conjunction of any number of true propositions is a true proposition, we may multiply each of these expressions together, and obtain the equation

$$(C_1 B'_2 + C'_1 B_2)(C_2 D'_3 + C'_2 D_3)(D_4 A'_2 + D'_4 A_2) = 1.$$

From this point onwards the working is automatic! As in elementary algebra, since the distributive law holds, the left-hand side may be multiplied out. We obtain eight terms, and write them all down:

$$C_1 B'_2 C_2 D'_3 D_4 A'_2 + C_1 B'_2 C'_2 D_3 D'_4 A_2 + C_1 B'_2 C'_2 D_3 D_4 A'_2$$
$$+ C_1 B'_2 C_2 D'_3 D'_4 A_2 + C'_1 B_2 C_2 D'_3 D_4 A'_2 + C'_1 B_2 C'_2 D_3 D'_4 A_2$$
$$+ C'_1 B_2 C'_2 D_3 D_4 A'_2 + C'_1 B_2 C_2 D'_3 D'_4 A_2 = 1.$$

Of these eight terms seven are 0 because they contain a symbol multiplied by its negation. Thus Cissie could not have come both first and second, and so the first and fourth terms are 0, each containing $C_1 C_2$. Again, since Doreen could not have been both third and fourth, the third and seventh terms are 0, each containing $D_3 D_4$. Finally, the sixth term is 0, because both Brenda and Alice could not have been second, and the fifth and eighth terms are both 0 because Brenda and Cissie could not both have been second, and we are therefore left with one term of the eight:

$$C_1 B'_2 C'_2 D_3 D'_4 A_2 = 1,$$

and none of the symbols in this term can be 0, so that each one is 1, and denotes a true proposition. This gives us, reading from the left:

"Cissie was first; Brenda was not second; Cissie was not second; Doreen was third; Doreen was not fourth; Alice was second."

This sounds very much like an oracular utterance, and it does indicate that Cissie was first, Alice was second, Doreen was third, and therefore Brenda was fourth.

The reader will agree that the problem solved is a more subtle one than either of the two Lewis Carroll ones in which, mathematically speaking, we merely had to arrange a number of quantities in order. There are many inferential problems which can be solved by Boolean algebra,* but we have probably said enough to indicate its utility. It has been used to disentangle the intricacies of insurance policies, and other complicated legal documents. It need hardly be pointed out that no use is made of it in finding the contradictions and *non sequiturs* in political speeches, for we do not use a complicated piece of machinery to detect the obvious.

Some logical paradoxes will be discussed in Chapter VIII. In our next chapter we shall leave the world of algebraic symbols for a time, and see how and why mathematicians have become interested in an important branch of geometry which should interest the makers of two-way-stretch garments also.

---

* T. J. Fletcher, *Math. Gazette*, Vol. 36 Sept. 1952.

CHAPTER V

TWO-WAY STRETCH

A SIMPLE geometrical figure, say a triangle, is drawn on the surface of a toy rubber balloon, and the balloon is then blown up. Is there any connection between the swollen figure with curved sides we now see and the original triangle? Distances have altered, and straight lines have become curved. But some properties remain. The new figure still has an outside and an inside, and a line drawn from the outside to the inside must still cross a boundary, made up of the three curved sides. These curved sides still intersect, in pairs, in

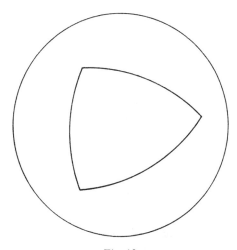

Fig. 10.

three points, which were the vertices of the original triangle. Points which were near to each other in the original figure still remain near to each other on the surface of the inflated balloon, provided that it does not burst. Every point on the surface of the balloon was originally a point in the original figure, and every point in the original figure becomes a point on the surface of the balloon.

The figure on the surface of the inflated balloon is called a *topological transformation* of the original figure, and *topology* is that branch of mathematics which seeks to determine the properties of geometrical figures which remain unchanged when the figure is subject to a topological transformation. It is tempting to call topology *rubber-sheet geometry*, and to hope that the manufacturers of two-way-stretch foundation garments will subsidise a chair for the study of this branch of mathematics. But we shall see that more general transformations than those afforded by stretching rubber sheets must be studied, and manufacturers of feminine underwear do not seem to need the help of the higher mathematics in their study of foundations.

Perhaps it is best to define a general topological transformation, having already given an example of a special topological transform-

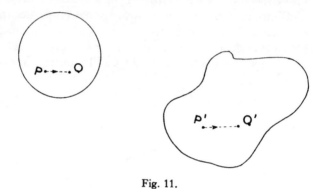

Fig. 11.

ation. We have two geometrical figures A and B, and regard B as a map of A. The two following properties must hold:

(1) *The mapping is one-to-one without exception.* This means that to every point P of the figure A there corresponds just one point P' of the figure B, and conversely.

(2) *The mapping is continuous in both directions.* This means that if we take any two points P, Q of the figure A, and move P so that the distance between it and Q approaches zero, then the distance between the mapped points P' and Q' of the figure B will also approach zero, and conversely.

The reader will have noticed that we call the figure B a *map* of the figure A, and he will wonder whether maps of the earth's surface are topological transformations. For limited areas this is indeed the case, but the *whole* surface of the earth cannot be mapped on to a plane by a topological transformation. Either condition (1) or condition (2) or both break down at one or more points on the earth's surface. Thus Mercator's projection breaks down completely at the poles.

The special topological transformations which arise in connection with figures drawn on rubber balloons or rubber sheets are called *deformations*. It is possible to have a topological transformation which cannot arise by deformation. Thus each of the two knots in Fig. 12 below is topologically equivalent to a circle. To show this, all that we need do is to cut each knot, untwist it, and sew the severed ends

Fig. 12.

together again, when we shall be able to make a circle with each piece of rope. Condition (1) will be satisfied after the ends are sewn together again, and also condition (2). It follows that each knot can be mapped topologically on the other. But neither can be deformed into a circle, or into the other.

If the reader is not tied in knots already, he may like to practise *the escape-artist's trick,* or *how to take off your waistcoat without taking your arms out of your jacket.* This comes within the orbit of topology, since both waistcoat and jacket are assumed to be deformable! Tearing is not permitted, even if subsequent stitching-up is possible.

Unbutton your jacket, and draw the back of your jacket over your head, keeping your arms in the sleeves, of course. Your jacket will then lie, much crumpled, across your chest.

Fig. 13.

Unbutton your waistcoat, and get it over the back of your head in a similar manner, so that it also lies across your chest.

Now get your jacket back over your head, into its normal position, so that only your waistcoat lies taut across your chest. Work one sleeve-hole of the waistcoat down your arm, and inside the jacket-sleeve, until you can slip it over the hand which passed through it originally. You will now find it easy to do the same thing with the other sleeve-hole of the waistcoat. The waistcoat can now be pulled away, down or up a jacket-sleeve, and you will have taken off your waistcoat without taking off your jacket.

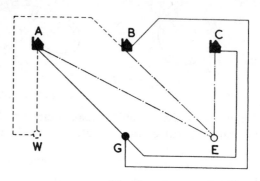

Fig. 14.

A pull-over may be used instead of a waistcoat. Dinner-jacket is probably the worst possible kind to use for a demonstration, but the author was present when an eminent and resourceful mathematician introduced a note of hilarity into the proceedings by substituting this trick for an after-dinner speech at a mathematical society dinner!

A pencil-and-paper problem, which involves no physical exertion, may also serve to amuse people at a party. Three houses, denoted by the three points *A*, *B*, and *C* in a plane, have each to be connected up with the water, gas and electricity-mains, represented by the points *W*, *G* and *E* in the plane. One requirement of the local authority is that the pipes and wires must not cross each other!

Experiment shows that this is impossible, but this need not be revealed until some time has been spent on the problem. It is a topological problem, since it is essentially unchanged by topological transformations.

The reader may feel that a given figure has too few topological properties to be of much interest. We therefore illustrate an important topological difference between figures in a plane. The two figures in

Fig. 15 are not topologically equivalent. The first of these consists of all the points interior to a circle, while the second consists of all the points contained between two concentric circles. Any closed curve lying in the domain *a* can be continuously deformed or shrunk down to a single point *within the domain*. A domain with this property is said to be *simply-connected*.

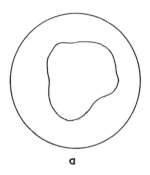   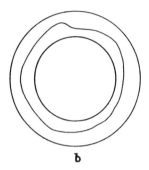

a   b

Fig. 15.

Now the domain *b* is not simply-connected. For example, a circle concentric with the two boundary circles and midway between them cannot be shrunk down to a single point within the domain since during this process the curve would necessarily pass over the centre of the two circles, which is not a point of the domain. The property of being simply-connected is evidently a topological property, so that our two figures are not topologically equivalent.

A domain which is not simply-connected is said to be *multiply connected*. If the multiply-connected domain we have just considered is cut along a radius, as in Fig. 16, the resulting domain is simply connected.

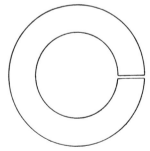

It is easy to construct figures with an even higher degree of connectivity. We merely increase the number of holes. Thus the domain in Fig. 17 is neither

Fig. 16.

topologically equivalent to the domain *a*, nor to the domain *b* of Fig. 15. In order to convert it into a simply-connected domain, two cuts are necessary.

If $n - 1$ cuts are necessary to convert a given multiply-connected domain D into a simply-connected domain, these cuts going from boundary to boundary, and not intersecting each other, then the domain D is said to be n-tuply connected. The degree of connectivity

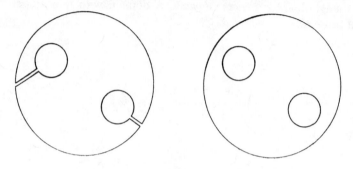

Fig. 17.

of a domain in the plane is an important example of a *topological invariant*, or *number* associated with a plane figure, this number being unchanged by topological transformations of the figure.

Two surfaces which appear a good deal in topology are the sphere and the torus. We do not need to describe the sphere. The torus, or

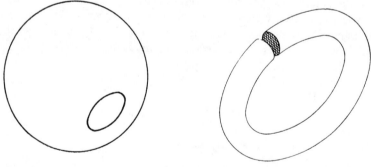

Fig. 18.

anchor-ring, is what an inflated bicycle-tube would look like if there were no valve. Doughnuts with a hole in them also approximate to the shape, as do quoits, and, of course, anchor-rings. We will show that a sphere is *not* topologically equivalent to a torus.

On the sphere, as in the plane, every simple closed curve (see Fig. 18) separates the surface into two parts. This means that if the surface of the sphere is cut along this curve, it will fall into two distinct and unconnected pieces; or what is the same thing, that we can find two points on the sphere such that any curve on the sphere which joins them must intersect the closed curve.

On the other hand, if the torus is cut along the closed curve shown, the resulting surface still hangs together. Any point on the surface can be joined to any other point by a curve which does not intersect the cut. This proves that the sphere and the torus are topologically distinct.

Now an ordinary surface has two sides. This applies both to closed surfaces like the sphere and the torus, and to surfaces with boundary curves, such as the torus from which a piece has been removed. The two sides of such a surface could be painted with different colours to distinguish them. If the surface is closed, the two colours never run into each other. If the surface has boundary curves, the two colours run into each other only along these curves. An insect crawling along such a surface, and prevented from crossing boundary curves, if any exist, would always remain on one side, and on the same colour.

Moebius (1790–1868) made the astonishing discovery that there are surfaces with only *one* side. The simplest such surface, appropriately called the *Moebius band*, is made by taking a long rectangular strip of paper, and pasting the two ends together, after giving one a half-twist, as in Fig. 19.

An insect crawling on this surface, keeping always to the middle of the strip, will return to its original position upside-down!

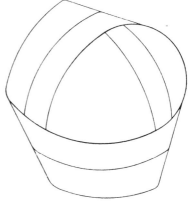

Fig. 19.

It is well-known that some window-cleaners will only do the insides of windows, whereas others only do outsides. The same applies to house-painters. But if a painter were set to work on the outside of the Moebius band, he would unwittingly have also painted the inside as well before he knew just what he was doing.

Moebius held the post of astronomer in a minor German observatory. At the age of sixty-eight he submitted a paper on "one-sided" surfaces to the Paris Academy. This contained some of the most surprising facts in the new science (as it was then) of topology. Like other important contributions before it, his paper lay buried for years in the files of the Academy, until it was eventually made public by the author. Moebius will remain famous as long as scissors and paper continue to exist on earth, for the Moebius band is a remarkable surface, and can afford much amusement. If it is cut along a centre line (shown in Fig. 19), it does not fall into two distinct strips of the same kind, as one might expect, but *remains in one piece!*

It is rare for anyone not already familiar with this surface to antici-pate this, because if one end had *not* been given a half-twist before being joined, so that we had a strip of paper like a napkin-ring, cutting along the central line *would* make it fall into two pieces. After cutting the Moebius band once, we can cut it again. It *then* falls into two distinct but intertwined pieces.

If the reader is enjoying himself with paper, gum and scissors, we should like to draw his attention to an easily constructed model which can provide even more pleasure than the Moebius band, although its theoretical importance is not so great. This model is of a ring of tetrahedra, which has the curious property that it can rotate inwards or outwards, like a smoke-ring. The effect is unexpected and remark-able, and delights anyone who handles it.

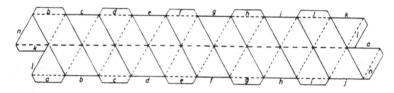

Fig. 20.

This model consists of ten tetrahedra, each joined to its neighbours by a pair of opposite edges to form a ring. The "net" for the model is given above. The triangles shown are all equilateral triangles. The model can be made with ordinary foolscap. The above net is drawn accurately on the foolscap, and then cut out. The paper is folded forwards along the lines marked on the net, and backwards along the dotted lines. The tabs shown are for sticking the model together. Tabs are joined to edges with the same letter.

A model which cannot be made with paper, but which presents a challenge to the glass-blower, is that of the Klein bottle. It is called after the great German mathematician Felix Klein (1849–1925) who made Goettingen famous as a centre for mathematical learning and research. This bottle, which has evidently inspired modern sculptors, is of importance as representing a one-sided surface which is closed, and has no boundary. It arises quite naturally from certain theoretical considerations, and was not invented, in the first place, for fun.

But once one is aware of the Klein bottle, one feels, as of so many mathematical constructs, that it has a certain something, quite indefinable of course, which other bottles have not got.

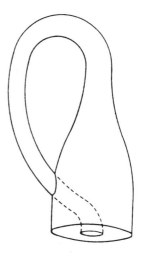

The question of the *boundary* of a surface, especially in the case of the Moebius band, may be troubling the reader. A surface is *closed*, and has no boundary, if all the points of the surface are *internal* points. A point of a surface is an *internal* point if a small circle can be drawn about it as centre *all internal points of which are points of the surface.*

All points of the Moebius band are internal points except those on the edge, or boundary. The sphere and torus are closed surfaces.

Fig. 21.

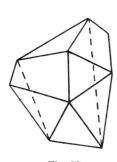

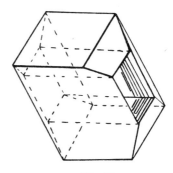

Fig. 22.

Fig. 23.

While topology is definitely a creation of the last hundred years, there are a few isolated earlier discoveries which later found a place in the modern systematic development. By far the most important of these is a formula which connects the numbers of vertices, edges and faces of a *simple* polyhedron. Now a polyhedron is a solid whose surface consists of a number of polygonal faces (triangles, quadrilaterals, pentagons, hexagons are all polygons). A cube is a polyhedron. So is a tetrahedron. In the case of the *regular* solids, all the polygons which make up the faces are congruent, and all the angles at vertices (corners) of the polyhedron are equal.

A polygon is *simple* if there are no "holes" in it, so that its surface can be deformed continuously into the surface of a sphere. Fig. 22 shows a simple polyhedron which is not regular, while Fig. 23 shows a polyhedron which is not simple.

Although the study of polyhedra held a central place in Greek geometry, it remained for Descartes and Euler to discover the following fact:

In a simple polyhedron let V denote the number of vertices, E the number of edges, and F the number of faces: then, *always*,

$$V - E + F = 2.$$

This relation is known as *Euler's formula*. It holds for the simple polyhedron of Fig. 22, where

$$V - E + F = 9 - 18 + 11 = 2,$$

but does not hold for the polyhedron of Fig. 23, which is not simple. Here we have

$$V - E + F = 16 - 32 + 16 = 0.$$

The reader should check that Euler's formula holds for all the simple regular polyhedra of Fig. 24.

The range of Euler's formula, however, goes far beyond the polyhedra of elementary geometry with their flat faces and straight edges. It applies equally well to a simple polyhedron with curved faces and edges, or to any subdivision of the surface of a sphere into regions bounded by curved arcs. In fact, if we imagine the surface of the polyhedron or of the sphere to be made of thin sheet rubber, the Euler formula will still hold if the surface is deformed by bending or stretching the rubber into any other shape, so long as the rubber is not torn in the process. For the formula is only concerned with the *numbers* of the vertices, edges and faces, and not with lengths, areas, straightness, or any other concepts of elementary geometry. The formula, in fact, is a topological formula.

We end this brief survey of a fascinating branch of modern mathematics with a description of *the four-colour theorem*, another of the great unsolved problems of mathematics.

In colouring a geographical map, it is customary to give different colours to any two countries with a common frontier. Now it has been found, as the result of long experience, that any map, no matter how

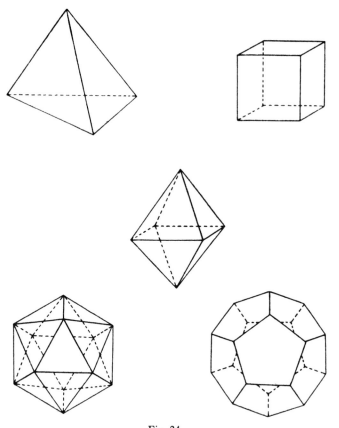

Fig. 24.

many countries it represents, nor how they are situated, can be so coloured by using only *four* distinct colours.

It is easy to see that no smaller number of colours will suffice for all cases. Fig. 25 shows an island that certainly cannot be properly coloured with less than four colours, since it contains four countries, each of which has a common boundary with the other three.

The fact that no map has yet been conceived whose colouring requires more than four colours suggests the following mathematical theorem:

For any subdivision of the plane into non-overlapping regions, it is always possible to mark the regions with the numbers 1, 2, 3, 4, each region being marked with one number, in such a way that no two adjacent regions have the same number.

By *adjacent* regions we mean regions with a whole segment of boundary in common. Two regions which meet at a single point, or at a finite number of points, will not be called adjacent, since no confusion would arise if they were treated with the same colour.

Many attempts have been made to prove this theorem, and the failure of the many has given mathematicians a great respect for the

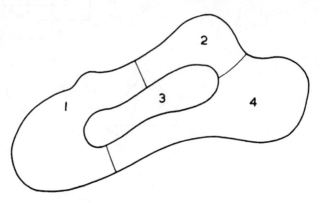

Fig. 25.

four-colour *hypothesis*. After all, unless it is proved, we do not know that the above statement is always true. But no counter-example, of a map for which more than four colours are necessary, has ever been devised. We shall not adopt the attitude of the poet who said:

> Would Mathematicals—forsooth—
> If true have failed to prove its truth?
> Would not they—if they could—submit
> Some overwhelming proofs of it?
> But still it totters *proofless*! Hence
> There's strong presumptive evidence
> None do—or can—such proof profound
> Because *the dogma is unsound*.
> For, were there means of doing so,
> They would have proved it long ago.

On the other hand it must be pointed out that the theorem cannot be proved by drawing a map and showing how it can be coloured. The proof must be a general one, and hold for any conceivable map. Every attempt has been made in this book, so far, to encourage the reader. But if he thinks he can find a simple proof of the four-colour theorem, let him reflect that some of the best brains in the world have failed to do so, and let him examine his proof carefully before he sends it for inspection!

Curiously enough, for surfaces more complicated than the plane, like the torus, the corresponding theorem has been proved. For the torus any map can be coloured using *seven* colours, and maps can be constructed for which this is the minimum number.

In our next chapter we leave geometry, and discuss the basis of all manipulation with symbols, algebra.

CHAPTER VI

RULES OF PLAY

IN this chapter we discuss the fundamental rules for the manipulation of symbols, "that more secret and subtill part of Arithmetike, commonly called Algebra." The word is of Arabic origin, *al-jebr* meaning the reunion of broken parts, and as such is applied to the surgical treatment of fractures. The Arabic mathematics was much concerned with the solution of quadratic equations, in which a term is *restored* to complete the square, and the equation is then *reduced* to the extraction of a square root. It seems fairly certain that the term *algebra* was meaningfully applied to this process. If to this derivation there is added the certainty that the term was fancifully identified with the name of the Arabic chemist Geber, the origin of the term is explained as well as any word can be explained. In any case, by *algebra* we mean a calculus of symbols combining according to certain defined laws.

It is found that certain mathematical systems satisfy the following laws:

Associated with every two elements a and b there is a unique element which we call the *sum*, and denote by $a + b$. The symbol $+$ obeys the law formulated by the equation

$$a + b = b + a.$$

Technically, we say that addition is *commutative*, meaning that a change in the *order* of addition does not affect the final result. Addition is also *associative*, by which we mean that

$$a + (b + c) = (a + b) + c,$$

in other words, the way we associate terms when adding does not affect the final result.

We define *zero*, written 0, as an element such that, for all a in the mathematical system we are considering,

$$a + 0 = 0 + a = a.$$

We can prove that if there is such an element, it must be unique. For if $0'$ were another element with the same properties, we should have, by definition,

$$0' + 0 = 0' = 0,$$

using the properties both of $0'$ and of 0.

If zero does exist, we define the *negative* $-a$ of an element a by the property

$$a + (-a) = 0.$$

The negative of an element a need not exist in the system, but if it does, it may also be shown to be unique. If negatives always exist, we find $-(-a)$ by finding a solution of the equation in x,

$$-a + x = 0.$$

Since the element a satisfies this equation, and since we know that the solution is unique, we see that

$$-(-a) = a.$$

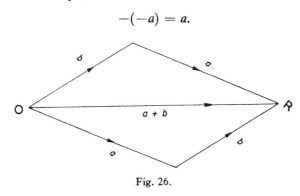

Fig. 26.

A mathematical system which obeys the above laws, in which zero exists, and in which every element has a negative—such a system is called an *additive group*. The point of the general discussion we have just given is that there are many such systems.

For example, the ordinary integers form an additive group, the term *addition* which is used above being ordinary addition. The reader may indignantly ask, "When is addition *not* ordinary addition?" Our next example illustrates the fact that it may be convenient to use the term addition for a process which is certainly not ordinary addition.

The elements we consider are *vectors*. For our purpose we shall define a vector as a straight-line displacement of a point. A vector therefore has magnitude and direction. We define the addition of vectors by considering the total displacement of the point if one displacement follows the other. To be precise, let a and b be the vectors shown in Fig. 26. Then $a + b$ is defined to be the vector $O R$. From the diagram it is clear that

$$a + b = b + a.$$

The zero vector is the vector of zero magnitude. Its direction is immaterial. The vector $-a$ has the same magnitude as a, but the opposite direction, and $a + (-a) = 0$, where 0 stands for the zero vector. Before the days of air-travel, vectors were thought to be very exotic mathematical plants. Every air-pilot uses them nowadays.

Another example of an additive group which is not the group of ordinary integers is given by the *rotations* of a plane about a fixed point. Take a plane, and mark a point O in it. Draw a line in the plane to pass through O. The elements of our mathematical system are to be the *rotations* of the plane about O. These rotations will be regarded as

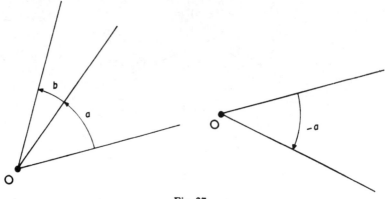

Fig. 27.

positive if obtained by an anti-clockwise rotation of the plane (the opposite direction to that of the hands of a clock). If a and b denote two rotations, and b takes place after a, the resultant, or total rotation, can be denoted by $a + b$. Since the total result is the same if b is carried out first, and a follows,

$$a + b = b + a.$$

The zero of this system represents no rotation of the plane, when the line drawn through O has not altered its position. If a represents any rotation, the line drawn through O returns to its original position if a is followed by an equivalent rotation in a *clockwise* direction. Thus the negative of an anti-clockwise rotation is a clockwise rotation through the same numerical angle.

The rotations of a plane about a fixed point therefore afford another example of an additive group. This group contains an infinite number of elements, as does the additive group of ordinary integers, but it is not difficult to give an example of an additive group containing a *finite* number of elements. Such a group is called a *finite* group.

All that we have to do is to restrict the rotations of the plane to multiples of the angle 360/n degrees, where n is any fixed integer. Then the only distinct positions of the fixed line in the plane drawn through O correspond to the division of 360 degrees into n parts. The diagram illustrates the case $n = 6$. The group contains n distinct elements.

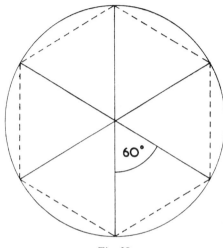

Fig. 28.

If a represents the smallest rotation in the group, not zero, then since a represents a rotation of 360/n degrees,

$$a + a + a + \ldots\ldots + a = 0,$$

where there are n terms in·the sum. For if the moving line returns to its original position, the resulting element of the group is denoted by 0.

There is no objection to writing $a + a = 2a$, $2a + a = 3a$, and so on, and the distinct elements of the group can therefore be written as

$$0, a, 2a, 3a, \ldots\ldots, (n - 1)a,$$

with

$$na = 0.$$

We have now extended the conventional notions of addition very considerably. A non-conventional use of the symbols 0 and 1 occurs very often in modern mathematics, and illustrations of this use will be found in a number of places in this book.

Ordinary integers, as we know, can be multiplied together, as well as added, and we now consider mathematical systems in which both operations are possible. We assume that, as far as addition goes, the elements form an additive group. We shall also assume that the operation of multiplication is such that the *product* of any two elements a and b is uniquely defined. We write this product as ab, but note that the product may depend upon the order in which the elements are taken: thus we do *not* assume that

$$ab = ba.$$

We *do* assume the *associative law of multiplication*:

$$a\,(b\ c) = (a\ b)\ c.$$

There must be a connection between addition and multiplication, and we assume the *distributive laws*:

$$a(b + c) = ab + ac; \quad (b + c)\,a = ba + ca.$$

These laws hold, of course, in ordinary arithmetic, and we have already come across them in the algebra of classes (Chapter IV). A mathematical system in which the elements satisfy the laws given above is called a *ring*. The ordinary integers form a ring, addition and multiplication being as ordinarily understood.

If we consider all the *even* integers, we see that they form, in the first place, an additive group, and, more extensively, that they form a ring, since addition and multiplication of even numbers by even numbers always produces even numbers. Before considering the operation of division, we deduce important consequences of the rules of addition and multiplication, which hold for the elements in any ring in which multiplication is *commutative*.

The distributive law enables us to raise expressions like $p + q$, called *binomial* expressions, to any power. Thus

$$
\begin{aligned}
(p + q)^2 &= (p + q)(p + q) \\
&= p(p + q) + q(p + q) \\
&= p^2 + pq + qp + q^2 \\
&= p^2 + pq + pq + q^2 \\
&= p^2 + 2pq + q^2. \\
(p + q)^3 &= (p + q)(p^2 + 2\,p\,q + q^2) \\
&= p^3 + 2{\cdot}p^2q + p\,q^2 + q\,p^2 + 2p\,q^2 + q^3 \\
&= p^3 + 3\,p^2\,q + 3\,p\,q^2 + q^3,
\end{aligned}
$$

and so on.

It is desirable to have a formula for the coefficients in the expansion of $(p + q)^n$, where n is any positive integer. Pascal, the great French seventeenth-century mathematician and writer (some people would invert the order here) noticed that the coefficients could be formed numerically by the use of an arithmetical triangle:

$$
\begin{array}{ccccccccccccc}
&&&&&& 1 &&&&&& \\
&&&&& 1 && 1 &&&&& \\
&&&& 1 && 2 && 1 &&&& \\
&&& 1 && 3 && 3 && 1 &&& \\
&& 1 && 4 && 6 && 4 && 1 && \\
& 1 && 5 && 10 && 10 && 5 && 1 & \\
1 && 6 && 15 && 20 && 15 && 6 && 1 \\
\end{array}
$$

. . . .

Each number of this Pascal triangle is the sum of the two numbers on either side of it in the preceding row. The second row gives the coefficients in $p + q$, the third row gives the coefficients in $(p + q)^2$, the fourth row in $(p + q)^3$ and so on. Thus the last printed row gives the coefficients in $(p + q)^6$:

$$(p + q)^6 = p^6 + 6p^5q + 15p^4q^2 + 20p^3q^3 + 15p^2q^4 + 6pq^5 + q^6.$$

To prove the theorem of the Pascal triangle, we suppose that

$$(p + q)^{n-1} = p^{n-1} + a_1p^{n-2}q + a_2p^{n-3}q^2 + \ldots + q^{n-1},$$

where the coefficients are $1, a_1, a_2, \ldots, 1$. Then

$$(p + q)^n = (p + q)(p^{n-1} + a_1p^{n-2}q + a_2p^{n-3}q^2 + \ldots + q^{n-1})$$
$$= p^n + (1 + a_1)p^{n-1}q + (a_1 + a_2)p^{n-2}q^2 + \ldots + q^n,$$

and we see that the coefficients are formed according to the Pascal triangle rule, each coefficient being the sum of the two coefficients on either side of it in the *preceding* row. Since the theorem is true when $n = 2$, it is true when $n = 3$, and therefore when $n = 4$, and so on.

If we want a formula for the coefficient of $p^{n-r}q^r$ in the expansion of $(p + q)^n$, we can prove that this coefficient is

$$\frac{n(n-1)(n-2) \ldots \ldots (n - r + 1)}{1.\, 2.\, 3. \ldots \ldots \ldots \ldots r}.$$

This is the content of the *Binomial Theorem*, which was discovered by Newton. It is convenient to call the continued product of r consecutive integers, beginning with 1 and ending with r, "factorial r", and to express this product by the symbol $r!$, so that we write

$$1.\, 2.\, 3. \ldots \ldots (r-1)(r) = r!.$$

This is sometimes facetiously read as "r shriek!", instead of "r factorial"!

With this notation, the coefficient above can be written

$$\frac{n(n-1)\,(n-2)\ldots\ldots(n-r+1)(n-r)(n-r-1)\ldots\ldots 3\,.\,2\,.\,1}{1\,.\,2\,.\,3\ldots\ldots\ldots\ldots\ldots r.\quad (n-r)(n-r-1)\ldots\ldots 3\,.\,2\,.\,1}$$

$$= \frac{n!}{r!(n-r)!}\,.$$

Before proving the binomial theorem, it is convenient to give an interpretation of the coefficients in the expansion of $(p+q)^n$. If we write

$$(p+q)^n = (p+q)(p+q)(p+q)\ldots\ldots\ldots(p+q)$$
$$= p^n + b_1 p^{n-1}q + b_2 p^{n-2}q^2 + \ldots + b_r p^{n-r}q^r + \ldots + q^n,$$

and consider how the terms arise in the product of the n terms $p+q$, we see that b_r is equal to *the number of ways in which r objects can be selected from amongst n distinct objects*. To see this, we first consider the product

$$(p_1 + q_1)(p_2 + q_2)\ldots\ldots(p_n + q_n),$$

and regard $p_1, p_2, \ldots, p_n$ as distinct objects. Then in the expansion of the product of these n terms, any *product* of r of the p's corresponds to a *selection* of r objects from amongst n distinct objects. The *number* of such selections gives the *coefficient* of $p^r q^{n-r}$ in the expansion of $(p+q)^n$.

Take the case $n = 3$. We have

$$(p_1 + q_1)(p_2 + q_2)(p_3 + q_3)$$
$$= p_1 p_2 p_3 + p_2 p_3 q_1 + p_3 p_1 q_2 + p_1 p_2 q_3 + p_1 q_2 q_3$$
$$+ p_2 q_3 q_1 + p_3 q_1 q_2 + q_1 q_2 q_3.$$

There are 3 ways of choosing two objects from amongst three distinct objects, corresponding to the terms $p_2 p_3 q_1$, $p_3 p_1 q_2$, and $p_1 p_2 q_3$. This is the same as the number of ways of choosing *one* object from amongst three. When we drop all the suffixes, we see that the coefficient of $p^2 q$ in $(p+q)^3$ is 3. Similarly in any other case.

We therefore write the coefficient of $p^r q^{n-r}$ in the expansion of $(p+q)^n$ as ^nC_r, this symbol representing the number of *combinations* of n objects taken r at a time. We now wish to prove that

$$^nC_r = \frac{n!}{r!(n-r)!}\,.$$

The theorem of the Pascal triangle states that

$$^nC_r = {}^{n-1}C_{r-1} + {}^{n-1}C_r.$$

If we assume that the formula we wish to prove is valid for all values up to $n-1$, we then have

$$^nC_r = \frac{(n-1)!}{(r-1)!(n-r)!} + \frac{(n-1)!}{r!\,(n-1-r)!}$$

$$= \frac{(n-1)!\,(r+n-r)}{r!\,(n-r)!} = \frac{n!}{r!\,(n-r)!}$$

Hence if the formula is true for all values up to $n-1$, we see that it is also true for n. Since the formula is true for $n-1 = 1$, when we have

$$^1C_0 = {}^1C_1 = 1,$$

our procedure shows that it is true for $n = 3$, and therefore for $n = 4, \ldots$ and so on. The theorem is therefore generally true.

This kind of proof is called *a proof by induction*. We have already come across it, above. It is of great importance in mathematics. More will be said about the method in Chapter VIII. To make our formula hold for the values $r = 0$, and also for $r = n$, we *define* $0! = 1$. We have already quietly assumed this definition when we wrote $^1C_0 = 1$, for we cannot say that nothing can be selected from one object in one way. We must use the formula, and put $0! = 1$, when we obtain the result.

Another proof of Newton's result can be obtained from the theory of *permutations*, or arrangements. We consider the number of arrangements of n distinct objects which are possible when the objects are arranged in order on a line. Take the case $n = 3$, and suppose that the objects are the letters p_1, p_2, and p_3. Then the distinct arrangements are

$$p_1 p_2 p_3, \quad p_1 p_3 p_2, \quad p_2 p_3 p_1, \quad p_2 p_1 p_3, \quad p_3 p_1 p_2, \quad p_3 p_2 p_1.$$

That this number is six can also be *deduced* from the following argument:

The first position on the line can be filled by one of the three letters. When this is done, the second position on the line can be filled by one of the two remaining letters. There are *six* ways, $6 = 3 \cdot 2$, therefore, of filling the first and second positions. The final position is automatically filled by the last remaining letter.

This argument can be applied to find the number of arrangements, or permutations, of n distinct objects on a line. The first position can be filled in n ways, by choosing any one of the n objects. When the first position is filled, the second position can be filled by choosing one of the remaining $n - 1$ objects. The first two positions can therefore be filled in $n(n - 1)$ ways. Proceeding thus, we see that the n positions on the line can be filled in $n(n-1)(n-2) \ldots \ldots 3 \cdot 2 \cdot 1 = n!$ ways.

Hence the number of permutations of n objects, taken n at a time, is $n!$ If, however, we wish to select only r objects from the n, and to arrange them on a line, the above argument shows that there are

$$n(n-1)(n-2) \ldots \ldots (n-r+1)$$

permutations. Now, in order to obtain these permutations, we could first *select* r objects from the n, and *then* rearrange these r objects on a line. Since r objects can be arranged in $r!$ ways, this gives the number of permutations of r objects selected from n as $r!$ (^nC_r), where ^nC_r is the number of ways of *selecting* r objects from n. Hence

$$r! \, (^nC_r) = n(n-1)(n-2) \ldots \ldots (n-r+1),$$

and we have proved again that

$$^nC_r = \frac{n!}{r! \, (n-r)!}$$

The term *permutation* is so well known nowadays, that it has been thought worth while to give the elementary theory. A knowledge of mathematics, as we have already hinted, would deter the rational man from filling in football coupons. But if he does concentrate on a small number of possibilities, a permutation will enable him to pay sixpence for each one, without overlooking any.

If in the binomial expansion we write $a = 1$, and $b = x$, the theorem we have just proved becomes

$$(1 + x)^n = 1 + nx + \frac{n(n-1) \, x^2}{2!} + \frac{n(n-1)(n-2) \, x^3}{3!} + \ldots \ldots ,$$

where, of course, the series on the right terminates because n is an integer. One of the amazing results which can be proved when we know enough about *infinite series* is that, provided x be suitably restricted, this expansion remains true even when n is *not* an integer. If n is not an integer, the series on the right never terminates, and we have an infinite series. We shall talk about infinite series, and what we mean by their *sum*, in Chapter VII. When we know what we mean by the sum of an infinite series, we can say that the sum of the infinite series on the right is equal to the function of x on the left.

We conclude our discussion of the basic rules of algebra by considering *division*. In some mathematical systems division, except by a possible unity, is not possible. For example, in the ring of ordinary integers we cannot carry out the operation of division by any integer, except 1, and obtain an integer in the ring. On the other hand, if we consider the system of *rational numbers*, that is the system of numbers of the form p/q, where both p and q are integers, we see that this system is a ring, and that in this ring division is always possible, except, of course, division by 0. For if p/q is an element of the system and $p \neq 0$, then q/p is also an element of the system, and

$$(p/q)(q/p) = 1.$$

The result of dividing 1 by p/q, or the *inverse* of p/q is therefore q/p, and so the result of dividing r/s by p/q is

$$(r/s)(q/p) = rq/sp.$$

A ring in which division by any non-zero element is always possible is called a *field*. The rational numbers therefore form a field.

We now consider division more systematically. If a is any element of a mathematical system, then the *inverse* of a in the system exists if there is a number a' in the system which is such that

$$a\,a' = a'\,a = 1,$$

where 1 is assumed to be an element of the system. If the element a' exists, then the result of dividing any element b of the system by a is $b\,a'$. We usually write the inverse of a as a^{-1}, so that

$$a\,a^{-1} = a^{-1}\,a = 1.$$

It is understood that 1 is the unity of the system, and is an element such that

$$a \cdot 1 = 1 \cdot a = a,$$

if a is any element of the system.

The reader may wonder why 0 is always excluded in division. Suppose that 0 did have an inverse. Let us call it $0'$. Then, by definition,

$$0 \cdot 0' = 1.$$

But we must have

$$0 \cdot b = 0,$$

where b is any element of the system, and in particular $0'$. This relation follows inevitably from

$$0 \cdot b = (a - a) \cdot b = a \cdot b - a \cdot b = 0,$$

and we must not have relations which contradict one another, at least, not in a mathematical theory.

There are many different kinds of mathematical fields, but we cannot discuss them here. It is worth mentioning, however, that *finite fields* exist, that is fields with only a finite number of elements. Carrying out the fundamental operations of addition, subtraction, multiplication and division on this finite number of elements always produces an element of the field. It is also possible to construct a set of elements which can be subjected to the basic operations, but are not commutative under multiplication, so that

$$a \cdot b \neq b \cdot a,$$

if a and b are suitable elements of the system. The *quaternions* are an example of such a system. They were invented by the great Irish mathematician William Rowan Hamilton (1805–1865). We have sacrificed a discussion of such systems to one on the application of groups to geometry. We shall discuss the *symmetry* of geometrical patterns. But before we do this, we must extend our notion of group. So far we have only considered commutative groups. These are usually called *Abelian* groups after the short-lived but illustrious Norwegian mathematician Niels Henrik Abel (1802–1829).

A system of distinct elements forms a *group* if the following four postulates are satisfied:

(1) *The group property.* To every ordered pair of equal or distinct elements of the system there is assigned a unique element of the system, the product of the two elements. We write $C = A B$, if the ordered pair of elements is A, B.

(2) *The associative law.* Products satisfy the equation $(AB)C = A(BC)$. The commutative law $AB = BA$ is not assumed.

(3) *The unit element.* There is an element E of the system which satisfies the following equations for every element A of the system:

$$AE = EA = A.$$

This element E is called the unity of the group.

(4) *The inverse element.* Every element A of the group has an inverse element $X = A^{-1}$ in the group which satisfies the equation $AX = E$.

In the case of an Abelian group, since $AB = BA$, always, we write the *product* of two elements as a *sum*, and the above postulates become those we have already considered for an *additive* group. This is merely a matter of convenience. Mathematicians are also human, and accustomed to thinking of a sum of two elements as being independent of the order of the elements. Hence, as a concession to human frailty, the product notation is changed to the sum notation, and there is less to keep in mind at any given instant. But groups which are not commutative exist. We take a simple example, which illustrates the generality of the definition of a group.

Consider the process of dressing oneself! The putting on of each garment may be considered as an element of a group. The inverse of putting on a particular garment is the taking of it off, and if the second follows the first, the total result is the identity, which represents leaving things as they are. Now if S represents the putting on of a shirt, and T represents the putting on of a tie, then ST represents the putting on of a shirt first and a tie second. The operation TS represents putting on a tie first and *then* a shirt, and the two results are not the same, so that $ST \neq TS$.

Another example is the group of motions of a plane over itself. We imagine a fixed plane, and a plane which can slide over it. Any motion of the moveable plane is an element of a group. The method of composition of elements S and T say, is to follow the motion S by the motion T, and to call the resulting motion ST. The unity element corresponds to no motion, and the inverse of any given motion is that motion which brings the plane back to its original position. To show that this group is not Abelian, take a point O in the fixed plane, and let S be a parallel translation of the plane which carries O into O_1. Let T be a rotation of the plane around O, of 90°. Then ST carries the point O into the point O_2, but TS carries O into the point O_1. Hence $ST \neq TS$, and the group is non-Abelian.

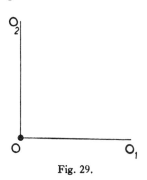

Fig. 29.

In order to investigate the symmetry of patterns, we must do more than consider the motions, or displacements of a plane over itself. We must also consider *reflections in a line*. If l is a given line in a plane, and P is any point in the plane, we drop a perpendicular PN on to the given line, and produce it to P', where $PN = NP'$. Then P' is the *reflection*, or geometrical image of P in the line l.

A reflection is a *transformation* or *mapping* of points of a plane which assigns a unique point P' to every point P of the plane. If we call this transformation T, then T is evidently *topological*, but what interests us more at the moment is that if we find the geometrical image of P', we arrive back at the original point P, so that we have

$$T(P') = P,$$
that is $$T(T(P)) = P,$$
or $$T^2 = E,$$

where E stands for the identity or unity transformation, which leaves every point of the plane unchanged.

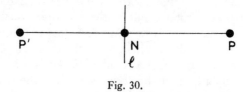

Fig. 30.

The simplest idea of symmetry in geometrical figures is that of *bilateral symmetry*. We observe this when there is a line in the figure such that *reflection in the line leaves the figure unaltered*. An example is shown below, taken from a capital in Saint-Denis d'Amboise. The artist has permitted himself a deviation from mathematical symmetry in some minor details.

Fig. 31.

Children are fond of making inkblots. Very interesting results can be obtained if the paper on which the blots are made is folded along a line while the blots are still very wet. This produces a mirror image of the blot in the line, and so a symmetric pattern is obtained. Such inkblots, in black and red ink, form the basis of a personality test much favoured by psychiatrists, the Rorschach test. The patient is handed a set of cards, each of which contains an inkblot, and the patient is asked to say what these blots suggest to him. Some patients see animals, trees, fairies, blood, and other things in these blots. Most

Fig. 32.

sane people become bored quite quickly, and cease to imagine that they can see anything more. The psychiatrist notes all this down, and with the help of the book of words, which goes with the cards, makes his deductions about the patient's personality from what the patient has said. This brief description of the procedure is not intended to praise or to deprecate the test. It is symmetrical inkblots which interest us at the moment.

Besides bilateral symmetry, there is rotational symmetry. We find this in figures which remain unchanged after rotation about some point. The circle is an obvious example, but we may also consider the mystic pentagram, so much favoured in mediaeval magic. This figure is carried into itself by the five anticlockwise rotations around

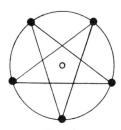

Fig. 33.

its centre O, the angles of which are multiples of 360/5 degrees, including the identity. It is also carried into itself by the five reflections in the lines joining O to the vertices, or corners, of the pentagon. These ten operations of rotation and reflection form a group, the group of self-transformations of the pentagram, and it is this group which expresses the mathematical symmetry of the figure.

If a given figure can be transformed into itself, it is easy to see that all the transformations which do this form a group, since a *sequence* of any two leaves the figure unaltered, and is therefore one of the *transformations* which leave the figure unaltered. The unity transformation does not change the position of any point of the figure, but maps every point on itself. The inverse of a self-transformation is easily defined, and is a self-transformation. All the group postulates are satisfied, and we have the *group of self-transformations of the figure*. The symmetry of the figure is expressed by this group.

Rotational symmetry is abundant in nature. Snow crystals provide the best-known specimens of hexagonal symmetry, of the sort

Fig. 34.

illustrated by the figure of a hexagon inscribed in a circle. These very beautiful patterns were widely displayed at the Festival of Britain, and were printed on textiles, wallpaper and pottery, but they were not a discovery of the architects of that Exhibition. They have been known from the early days of micro-photography.

We now consider one-dimensional patterns or "ornaments". There are two types. The first is that produced by simple movement, or translation. Any pattern in the plane displaced in a regular way gives a spatial rhythm. For example, the continued application of a stencil, a device familiar to the amateur interior-decorator, produces such a pattern, as we see below.

There is another kind of symmetry on a line. We can reflect the points of a line in one of its points O. Reflection in O, followed by the

Fig. 35.

translation or movement OA is equivalent to reflection in the point A_1 which bisects OA.

If this is not evident from the diagram it is easily proved. The two types of symmetry we have just described are the only possible ones

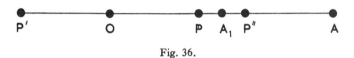

Fig. 36.

on a line. In the second type, if the basic translation be through a distance a, the centres of reflection, by the result just mentioned, follow each other at half the distance $a/2$. The crosses in the diagram below mark the centres of reflection.

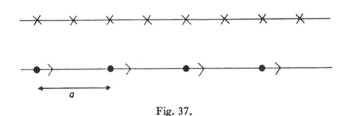

Fig. 37.

A very frequent motif in Greek art, the palmette, illustrated below is of the reflection plus translation type.

Fig. 38.

So much for one-dimensional symmetry. A more complicated kind of symmetry is illustrated in wall-papers, carpets, tiled floors and dress-materials. To discuss this *two-dimensional* symmetry, we intro-

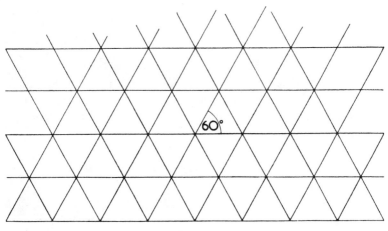

Fig. 39.

duce the idea of a *lattice* of points in a plane. We take two coordinate axes, inclined at an angle a to each other, and we plot the points (ma, na), where a is a fixed number, and m and n are positive or negative integers. Joining up these points, by lines parallel to the x

and y axes, we obtain the mesh shown above. In this diagram we have taken $\alpha = 60°$, and also drawn the diagonals of the parallelograms in the figure.

The symmetry of the lattice, or of any pattern based on it, is expressed mathematically by the group of transformations of the plane which leave the pattern unchanged. The lattice itself is unchanged, for any value of α, by motions of the plane parallel to the

Fig. 40.

x or y axes through distances ka, where k is a positive or negative integer. If we *rotate* the plane about one of the lattice points through a suitable angle, the set of displaced points will fall on the original points, which is what we want, if and only if the angle α is $360/n$ degrees, where n can only take the values 3, 4 or 6.

This result can be proved practically, of course, using a sheet of transparent paper on which the lattice is traced, rotating the tracing over the lattice about one of the points of the lattice, and seeing in what circumstances the points of the tracing cover the points of the lattice again.

When $n = 4$, so that $a = 90°$, we obtain the pattern of a floor covered with square tiles. When $n = 3$, or $n = 6$, so that $a = 120°$ or $a = 60°$, we obtain the lattice already drawn in Fig. 39, which is a pattern of hexagonal or triangular tiles, according to choice.

A beautiful example of a pattern derived from the hexagonal lattice is afforded by the illustration on page 107 of the fourteenth-century window of a mosque in Cairo. It is based on a trefoil knot, the various units of which are interlaced with superb artistry.

The depth of geometric imagination and inventiveness reflected in the design of this and similar patterns can hardly be overestimated. In fact the art of ornament contains in implicit form the oldest piece of higher mathematics known to us. The conceptual means for a complete abstract formulation of the problem, namely the mathematical notion of a group of transformations, did not arise until the nineteenth century. We know now that the 17 types of symmetry which can be found in the work of Arabic craftsmen exhaust all the possibilities. The mathematical proof of this result was only given in 1924, by George Polya.

We have not been able to do more than to hint at the mathematics which underlies the theory of ornament. Just as there are gifted people who can compose fugues without any understanding of the theory of the art of fugue, so there have always been artists who instinctively understand the laws of symmetry. Analysis limps a long way behind.

It may be mentioned here that the first attempts have been made to give a mathematical foundation to the theory of aesthetics. The interested reader will find much to ponder on in *Aesthetic Measure*, by G. D. Birkhoff, Harvard University Press, 1933. The author was a very distinguished mathematician.

CHAPTER VII

AN ACCOUNTANT'S NIGHTMARE

W E have all, at some time or another, added up columns of figures, and heaved a sigh of relief when we came to the end. What should we do if the columns never terminated, and were of infinite length? Would our sum, as we went down a column, grow larger without bound, or might it happen that at some stage we could prophesy that the sum would never exceed a definite fixed number? We discuss this question here, and apply our procedures to infinite decimals. We also consider some of the fundamental numbers of mathematics which are non-terminating decimals, and even mention that curious race of eccentrics, found all over the world, who have but one ambition, to prove that mathematics is wrong!

We begin with a simple sum of a finite number of terms,

$$1 + 2 + 2^2 + 2^3 + \ldots + 2^{n-1},$$

and obtain a formula for this sum of n terms. We notice that

$$1 + 2 = 3 = 2^2 - 1,$$
$$1 + 2 + 2^2 = 7 = 2^3 - 1,$$
$$1 + 2 + 2^2 + 2^3 = 15 = 2^4 - 1,$$

and feel tempted to apply the method of induction. We therefore *assume* that

$$1 + 2 + 2^2 + \ldots + 2^{n-2} = 2^{n-1} - 1.$$

It then follows that

$$1 + 2 + 2^2 + \ldots + 2^{n-2} + 2^{n-1}$$
$$= 2^{n-1} - 1 + 2^{n-1}$$
$$= 2 \cdot 2^{n-1} - 1$$
$$= 2^n - 1.$$

The Principle of Induction now assures us that

$$1 + 2 + 2^2 + \ldots + 2^{n-1} = 2^n - 1,$$

for all values of n. If a direct method of obtaining this result is preferred, and we call the sum of the terms S_n, the symbol standing for "the sum to n terms", we have

$$1 + 2 + 2^2 + \ldots + 2^{n-1} = S_n,$$
$$2 + 2^2 + \ldots + 2^{n-1} + 2^n = 2S_n,$$

so that, on subtraction, $2 S_n - S_n = S_n = 2^n - 1$.

As we increase n, this sum obviously becomes larger and larger. The reader has probably heard of the Emperor who was unable to form a rapid estimate of the rate of growth of this sum. The game of chess was invented during this Emperor's reign, and so delighted the Emperor that he sent for the inventor, and offered to reward him. The inventor asked whether he could be given a grain of rice for the first square on a chessboard, two grains for the second square, four grains for the third square, eight grains for the fourth square, and so on. This seemed a trivial reward to the Emperor, and he readily agreed. Now a chessboard contains 64 squares, so that the number of grains of rice asked for was

$$1 + 2 + 2^2 + \ldots + 2^{63} = 2^{64} - 1.$$

A simple approximation would have warned the Emperor that he was promising something beyond the capacity of his granaries to fulfil. For $2^4 = 16$, which is certainly greater than 10, so that

$$2^{64} = (2^4)^{16} > (10)^{16}.$$

The number of rice grains is therefore a number of at least 17 digits, and thus greater than a thousand million million. Actually

$$2^{64} - 1 = 18, 446, 744, 073, 709, 551, 615,$$

and contains twenty digits. Our approximation was a rough one, but not a useless one. The inventor was asking for more rice than has ever existed at any one time on this earth.

Very few people would be caught out nowadays in this way. But there are sums which the average man would disregard as trivial, but are far from insignificant.

Not very long ago there lived a rich merchant who was a citizen of a small country with a large national debt. Although the British had left, oil had not yet been discovered, and there was nobody to blame for its absence. One day this merchant wrote to the Prime Minister as follows:

"Dear Prime Minister,

Being filled with the most altruistic motives, as is highly natural for citizens of our beloved country, I make the following offer:

Beginning on December 10th next, which is my birthday, I shall convey to our Exchequer, free of all expense, £1 on the first day, £$\frac{1}{2}$ (one half) on the second day, £$\frac{1}{3}$ (one third) on the third day, and so on indefinitely, until our National Debt is wiped out. This offer, if accepted, will help to preserve the independence of our beloved country, so recently freed from the Imperialist yoke."

There followed some remarks of a more personal nature, which do not concern us here. Now the Prime Minister, who was no mathematician, was not very impressed with this offer. He knew the merchant well, and suspected a snag somewhere. But he sent for the Minister of Finance, and showed him the letter. As the country we are discussing was a modern state, newly created, with no traditions worth mentioning, the Minister of Finance knew some mathematics, and even something about decimal points.

"There is more in this offer than meets the eye", said the Minister of Finance. "It is true that on the surface it does not impress, but I think I can show you that it is worth considering." And this is what the Finance Minister demonstrated:

We have to sum, in pounds, the series

$$1 + \tfrac{1}{2} + \tfrac{1}{3} + \tfrac{1}{4} + \tfrac{1}{5} + \tfrac{1}{6} + \tfrac{1}{7} + \ldots$$

We group these fractions in the following way:

$$1 + (\tfrac{1}{2} + \tfrac{1}{3}) + (\tfrac{1}{4} + \tfrac{1}{5} + \tfrac{1}{6} + \tfrac{1}{7})$$
$$+ (\tfrac{1}{8} + \tfrac{1}{9} + \tfrac{1}{10} + \tfrac{1}{11} + \tfrac{1}{12} + \tfrac{1}{13} + \tfrac{1}{14} + \tfrac{1}{15})$$
$$+ (\tfrac{1}{16} + \ldots\ldots + \tfrac{1}{32}) + \ldots\ldots\ldots,$$

so that the first bracket contains two terms, the second bracket contains four terms, the third bracket contains eight terms, the fourth bracket sixteen terms, and so on. We now approximate very roughly, as follows, knowing that one-half is greater than one quarter, one third is greater than one quarter: in symbols

$$\tfrac{1}{2} > \tfrac{1}{4}, \qquad \tfrac{1}{3} > \tfrac{1}{4},$$

so that the sum of the two terms in the first bracket exceeds

$$\tfrac{1}{4} + \tfrac{1}{4} = \tfrac{1}{2}.$$

Again $\tfrac{1}{4} > \tfrac{1}{8}, \qquad \tfrac{1}{5} > \tfrac{1}{8}, \qquad \tfrac{1}{6} > \tfrac{1}{8}, \qquad \tfrac{1}{7} > \tfrac{1}{8},$

so that the sum of the four terms in the second bracket exceeds

$$\tfrac{1}{8} + \tfrac{1}{8} + \tfrac{1}{8} + \tfrac{1}{8} = \tfrac{1}{2}.$$

The sum of the eight terms in the third bracket, by the same procedure, exceeds

$$8(\tfrac{1}{16}) = \tfrac{1}{2},$$

and the sum of the sixteen terms in the fourth bracket exceeds

$$16(\tfrac{1}{32}) = \tfrac{1}{2},$$

and so on. Continuing in this way, we see that the sum offered exceeds

$$1 + \tfrac{1}{2} + \tfrac{1}{2} + \tfrac{1}{2} + \tfrac{1}{2} + \ldots$$

pounds.

"Ha!" said the Prime Minister, "this demonstration puts quite a different complexion on the matter. Even a mere Arts graduate like myself can see that a sufficient number of halves can make up any number of pounds, and so wipe our National Debt off the surface of the map." And the Prime Minister began to have very pleasant thoughts of a trip to Paris, London and New York, for business reasons, of course. "We must accept this offer!" he declared, "it is obviously in the national interest!"

But a horrid afterthought had struck the Finance Minister, and he had been frantically doing sums on a scrap of paper, in the best mathematical tradition, while the Prime Minister had been concerned with less abstract matters. Suddenly the Finance Minister tore the paper into little pieces, and exploded: "This offer is of no use at all!" he shouted, "We shall all be dead long before the sum paid over amounts to anything substantial. National Debt indeed! Let me show you, sir, what a mean offer this really is." And this is what he showed the Prime Minister:

On the first day, £1 is paid over. To obtain a further sum greater than £$\tfrac{1}{2}$, two more days must elapse. Then to make sure of a payment greater than another £$\tfrac{1}{2}$, four more days must elapse; then eight days, then sixteen days, and so on. In fact, to make certain of a payment greater than

$$\pounds(1 + \tfrac{1}{2} + \tfrac{1}{2} + \ldots \ldots + \tfrac{1}{2}),$$

where there are n halves added together, the number of days over which payment is spread is

$$1 + 2 + 4 + 8 + \ldots + 2^n,$$

and this sum, as we saw above, is

$$2^{n+1} - 1.$$

Hence, to make sure that even £10,000 was paid over, the Exchequer would have to wait more than 2^n days, where $n = 20,000$. Since, as we saw above, 2^{63} is an astronomically large number, and this number of days is far, far larger, the Finance Minister had good reason to be depressed. The offer was not accepted.

The series $1 + \frac{1}{2} + \frac{1}{3} + \frac{1}{4} + \ldots + 1/n + \ldots$ we have just considered is called the *harmonic series*, and we say that it *diverges to infinity*, but very slowly. It is clear that in mathematics we neglect small quantities at our peril. The cumulative effect can be overwhelming.

Fortunately not all series become infinitely big as we add up the terms. A well-behaved series is the *geometric series*,

$$1 + r + r^2 + r^3 + \ldots + r^{n-1} + \ldots$$

if r is numerically less than 1. We sum the series to n terms, and let

$$1 + r + r^2 + \ldots + r^{n-2} + r^{n-1} = S_n.$$

Then $$r + r^2 + \ldots + r^{n-1} + r^n = rS_n,$$

and on subtraction we find that

$$1 - r^n = (1 - r) S_n,$$

so that we have a formula for the sum to n terms,

$$S_n = \frac{1 - r^n}{1 - r}.$$

Now S_n can be split into two parts, thus:

$$S_n = \frac{1}{1 - r} - \frac{r^n}{1 - r}.$$

The first part is fixed, and independent of n. If r is positive and less than 1, then as n increases without bound, r^n steadily diminishes to zero. This is also true when r is negative and numerically less than 1, but we are more interested in the positive case. Hence, if r is thus restricted, the sum of n terms of the series approaches steadily to the fixed number $1/(1 - r)$ as n increases, and can be made to differ from this fixed number by as small a quantity as we please if we make n large enough. We must add to this the important fact that if this difference be small enough when $n = n_0$, it remains small enough when $n > n_0$. We shall return to this condition in a moment.

We now say that the infinite series

$$1 + r + r^2 + \ldots + r^{n-1} + \ldots$$

has a *sum*, or is *convergent*, and that its sum is $1/(1-r)$. We shall make use of this series when we study infinite decimal expansions. But first let us give an example in which the sum of an infinite geometric series is intuitive.

A segment of length 2 units is drawn, and half of it is rubbed out. The remaining segment of length 1 is bisected, and one half rubbed out. The remaining segment of length $\frac{1}{2}$ is bisected, and one half is rubbed out. We can proceed thus indefinitely. The length rubbed out is

$$1 + \tfrac{1}{2} + \tfrac{1}{4} + \tfrac{1}{8} + \ldots\ldots$$

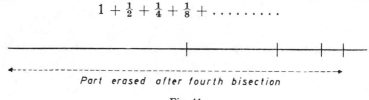

Part erased after fourth bisection

Fig. 41.

and we see from the diagram that this approaches the whole segment more and more nearly as we continue with our erasing process. Hence

$$1 + 1/2 + 1/(2)^2 + 1/(2)^3 + \ldots$$

is a convergent infinite series with the sum 2. This tallies with the result above, with $r = \frac{1}{2}$.

If we consider the infinite series

$$1 - 1 + 1 - 1 + 1 - 1 + \ldots\ldots,$$

we see that the sum of the first two, four, six, eight, $\ldots$, $2n$ terms is 0, but that the sum of the first one, three, five, seven, $\ldots$ $2n - 1$ terms is 1. For an infinite number of values of n therefore, in fact all the even values, the sum is 0, and for an infinite number of values of n, all the odd values, the sum is 1. But there is no number S such that the sum of n terms of the series is as near to S as we please *as soon as n exceeds some definite number n_0*. Hence we do not assign a sum to such an infinite series, but say that it *oscillates*.

Some interesting examples of the troubles which may arise when infinite series are not handled with care will be given later. But we now apply our summation of an infinite geometric series to infinite decimals.

We remarked in Chapter I that ordinary numbers or integers are expressed in the scale of ten. When we continue the process for numbers less than 1 and greater than 0, we obtain the ordinary decimal system. For example, the decimal 0.25 stands for $2/10 + 5/(10)^2$, and 0.217 represents $2/10 + 1/(10)^2 + 7/(10)^3$, and so on.

But we find, if we try to determine the decimal corresponding to certain fractions, such as $\frac{1}{3}$, that we obtain an infinite decimal expansion,

$$0.333333333 \ldots \ldots \ldots ,$$

in which the digit 3 is said to *recur*. Some fractions yield a *sequence* of recurring figures. For instance

$$\tfrac{1}{7} = 0.142857\ 142857\ 142857 \ldots \ldots ,$$

with the sequence 142857 recurring. These last two examples differ from the first two in being *infinite* decimals. Each of the infinite decimals is an example of an infinite geometric series, as we now show.

The recurring decimal

$$.333333 \ldots \ldots = 3/10 + 3/(10)^2 + 3/(10)^3 + \ldots .$$
$$= \tfrac{3}{10} (1 + 1/10 + 1/(10)^2 + \ldots .),$$

and the infinite geometric series within the brackets has the sum

$$\frac{1}{1 - 1/10} = \frac{10}{9} ,$$

so that the infinite decimal $= (3/10)(10/9) = \frac{1}{3}$, as it should.

In the second case, the recurring decimal stands for

$$\frac{142857}{(10)^6} \left(1 + \frac{1}{10^6} + \frac{1}{10^{12}} + \ldots \ldots \ldots \right) ,$$

and the infinite geometric series within the bracket has the sum

$$\frac{1}{1 - 1/10^6} ,$$

so that the whole decimal is equal to the fraction

$$\frac{142857\ (10)^6}{999999\ (10)^6} = \frac{1}{7} .$$

These two examples are rather special in that the decimal contains no non-recurring integers. If we consider the decimal .217 13131313

. . . . , where the digits 13 recur, we see that this decimal represents

$$2/10 + 1/10^2 + 7/10^3 + 1/10^4 + 3/10^5 + 1/10^6 + \ldots$$

$$= \frac{217}{1000} + \frac{13}{10^5} \left(1 + \frac{1}{10^2} + \cdots \cdots \cdots \right)$$

$$= \frac{217}{1000} + \frac{13}{10^5} \left(\frac{1}{1 - 1/10^2} \right) = \frac{2687}{12375}.$$

It is now clear that any recurring decimal is equal to a proper fraction p/q, where p and q are integers, and $p < q$. It is important to prove the converse of this result. We wish to show that if we find the decimal form of a proper fraction p/q, we obtain either a *finite* decimal, or at worst a *recurring* decimal. That is, we wish to show that the only kind of infinite decimal equal to a rational number is of the kind in which a sequence of numbers eventually recurs. The following proof is fairly mathematical, and can be assumed by the reader if he is merely interested in the result.

Let $r = p/q$, and let us suppose in the first instance that q is prime to 10, which means that q and 10 have no factors in common. We consider the various powers of 10:

$$10, 10^2, 10^3, 10^4, \ldots \ldots$$

and the remainders when each of these powers of 10 is divided by q. These remainders are all less than q, and can only be, in some order,

$$1, 2, 3, \ldots \ldots \ldots, q - 1.$$

Since the number of powers of 10 is infinite, we must be able to find two integers n_1 and n_2 such that 10^{n_1} and 10^{n_2} give the *same remainder* on division by q.

Assume that $n_1 > n_2$. Then the number

$$10^{n_1} - 10^{n_2} = 10^{n_2} (10^{n_1 - n_2} - 1)$$

is divisible by q, and since q is prime to 10, and therefore to powers of 10, it follows that $10^n - 1$ is divisible by q, where $n = n_1 - n_2$.

Let $(10^n - 1)/q = s$. Then

$$r = \frac{p}{q} = \frac{ps}{10^n - 1} = \frac{ps}{10^n} \left(\frac{1}{1 - 1/10^n} \right).$$

We know that

$$1/10^n + 1/10^{2n} + \ldots \ldots = \frac{1}{10^n(1 - 1/10^n)},$$

so that, writing $ps = P$, we have

$$r = \frac{P}{10^n} + \frac{P}{10^{2n}} + \cdots\cdots\cdots$$

Hence in this case $r = p/q$ is a pure recurring decimal, with n recurring figures.

We test this procedure with one of the pure recurring decimals already found, that corresponding to $\frac{1}{7}$. If we write down successive powers of 10, the remainders on division by 7 are:

$$10, \text{ remainder } 3,$$
$$10^2, \quad\text{,,}\quad 2,$$
$$10^3, \quad\text{,,}\quad 6,$$
$$10^4, \quad\text{,,}\quad 4,$$
$$10^5, \quad\text{,,}\quad 5,$$
$$10^6, \quad\text{,,}\quad 1,$$
$$10^7, \quad\text{,,}\quad 3.$$

Hence 10 and 10^7 give the same remainder after division by 7, and

$$10^7 - 10 = 10(10^6 - 1)$$

is divisible by 7. Since 10 is not divisible by 7, it follows that

$$10^6 - 1 = 999999$$

is divisible by 7, and in fact

$$\frac{10^6 - 1}{7} = 142857.$$

Hence $\dfrac{1}{7} = \dfrac{142857}{10^6 - 1} = \dfrac{142857}{10^6(1 - 1/10^6)}$

$$= \frac{142857}{10^6} (1 + 1/10^6 + 1/10^{12} + \cdots\cdots).$$

Hence once again we have shown that $\frac{1}{7}$ is a pure recurring decimal with 6 recurring digits, and the working should help to elucidate points in the proof which may have seemed difficult at first.

There is now the other case to consider in evaluating $r = p/q$, the case in which q is not prime to 10. This means that q contains 2 or 5 as factors. Take out all the possible factors 2 or 5. We may then write

$$q = 2^a 5^b Q,$$

where Q is prime to 10. Let m be the greater of a and b. Then

$$10^m(r) = 10^m(p/q) = 10^m \left(\frac{p}{2^a \, 5^b \, Q} \right)$$

has, on simplification, a denominator prime to 10. By what we have proved above, $10^m(r)$ is therefore expressible as the sum of an integer and a pure recurring decimal. But this is not true for $10^n(r)$, for any value of n less than m. It therefore follows that the decimal expansion of r has exactly m non-recurring digits, and these are followed by a sequence of recurring digits.

We illustrate this proof by considering the decimal for

$$r = \frac{1}{60} = \frac{1}{2^2 \times 5 \times 3}.$$

Here $m = 2$, and $10^2(r) = 5/3 = 1.666\ldots\ldots$, so that $r = .01666$ ... and has only two non-recurring figures.

We have now proved that any *rational* number, that is any number which can be expressed in the form p/q, where both p and q are integers, can be expressed as a finite or as an infinite recurring decimal. This decimal need not be a pure recurring decimal, but after a finite number of digits the decimal must recur.

Conversely we have also seen that any finite or recurring decimal, not necessarily pure, is equal to a rational number. Now, any finite decimal can also be written as an infinite, and necessarily recurring decimal. For the recurring decimal $.99999\ldots\ldots\ldots$, when summed, is equal to

$$9/10 + 9/10^2 + \ldots\ldots\ldots = \frac{9}{10} \left(\frac{1}{1 - 1/10} \right) = 1.$$

Hence we may write $.217 = .216999\ldots\ldots$, and similarly every finite decimal may be written as an infinite recurring decimal.

We must now show that the expression of a number as an *infinite* decimal is unique. We have used this theorem already in Chapter III, p. 62. Let us suppose, in fact, that

$$\frac{r^{(1)}}{10} + \frac{r^{(2)}}{10^2} + \frac{r^{(3)}}{10^3} + \cdots = \frac{s^{(1)}}{10} + \frac{s^{(2)}}{10^2} + \frac{s^{(3)}}{10^3} + \cdots,$$

where the $r^{(i)}$ and $s^{(j)}$ are all integers between 0 and 9. If $r^{(n)} = s^{(n)}$ for all values of the index n, there is nothing to prove. Let us suppose, then, that for $n = m$, $r^{(m)} < s^{(m)}$, and that for $n < m$, $r^{(n)} = s^{(n)}$.

Then

$$\frac{r^{(m)}}{10^m} + \frac{r^{(m+1)}}{10^{m+1}} + \frac{r^{(m+2)}}{10^{m+2}} + \cdots\cdots$$

$$\leqq \frac{r^{(m)}}{10^m} + 9\left(\frac{1}{10^{m+1}} + \frac{1}{10^{m+2}} + \cdots\right)$$

$$= \frac{r^{(m)}}{10^m} + \frac{1}{10^m} = \frac{r^{(m)} + 1}{10^m} \leqq \frac{s^{(m)}}{10^m}$$

$$\leqq \frac{s^{(m)}}{10^m} + \frac{s^{(m+1)}}{10^{m+1}} + \cdots\cdots\cdots$$

(All that we have done in the second term of this inequality is to sum the geometric series). But the first and final terms of the inequality are equal, by hypothesis. It follows that all the $\leqq$ signs must be replaced by $=$ signs, and therefore $r^{(n)} = 9$, and $s^{(n)} = 0$ for $n > m$.

We therefore return to the case above. But if two *infinite* decimals are equal, we have proved that every digit in the one expansion must coincide with the corresponding digit in the other.

This proves all the fundamental results on decimal expansions we shall need. A problem of the greatest theoretical importance now demands attention. What can we say about an infinite decimal which does not recur?

It is easy to write such a decimal down. Consider

$$.101001000100001000001\ldots\ldots$$

in which the number of zeros between the successive 1's increases by one at each stage. This is certainly a non-recurring decimal. Hence it is not equal to a rational number. We call such numbers *irrational*. We have already seen, in Chapter I, p. 33, that the square root of 2 is not a rational number. It follows that in the decimal expansion of this root, there are no recurring sequences of digits.

Many of the fundamental numbers which appear in mathematics are irrational numbers. One such is the ratio of the circumference of a circle to its diameter, commonly known by the Greek letter π.

The irrationality of π was proved by Lambert in 1761. It can also be proved that it is impossible to *square the circle*, that is to construct, by geometrical means, a square whose area is equal to the area of a given circle. But this mathematical proof has not deterred the circle-squarers, and many attempts have been made to show that the mathematicians are wrong. Before we discuss one of the most amusing of these attempts, let us see what mathematical expressions exist for π.

It can be proved, by using the integral calculus, that

$$\pi/4 = 1 - \tfrac{1}{3} + \tfrac{1}{5} - \tfrac{1}{7} + \tfrac{1}{9} - \tfrac{1}{11} + \cdots,$$

so that we have an infinite series by which π can be evaluated to any number of decimal places. This series *converges slowly*, which means that we should have to sum a large number of terms to ensure accuracy to, say, ten places of decimals. A far better series, which converges more rapidly, is

$$\pi/4 = 4(1/5 - 1/(3.5^3) + 1/(5.5^5) - 1/(7.5^7) + \ldots)$$
$$- (1/239 - 1/(3.239^3) + 1/(5.239^5) - 1/(7.239^7) + \ldots).$$

This series was discovered by Machin, in the early eighteenth century. A representation of π as an infinite product of rationals is due to John Wallis (1616–1703). He found that

$$\frac{\pi}{2} = \frac{2 . 2 . 4 . 4 . 6 . 6 . 8 . 8 \ldots\ldots}{1 . 3 . 3 . 5 . 5 . 7 . 7 . 9 . 9 \ldots\ldots}$$

If we take only the first digits in numerator and denominator, we obtain $\pi = 4$, which is not a good approximation! If we take the first two digits in numerator and denominator, we obtain $\pi = 8/3$, which is too small. If we take the first three digits in numerator and denominator, we find that $\pi = 32/9$, which is too large. The first four digits above and below give $\pi = 128/45$, which is too small. As we go on, the rational numbers which approximate to π move in towards each other, their ultimate value of convergence, which, of course, they never reach, π not being a rational number, being the value of π.

In the nineteenth century Shanks calculated the expansion of π to more than six hundred decimal places. The expansion begins 3.14159 . . . . . This expansion, when typed out, fills one side of a sheet of foolscap, if we go as far as Shanks went. A favourite lecturing trick of a well-known mathematician, still happily with us, is to write the Shanks approximation on a blackboard from memory, both forwards and backwards, his audience having been given typed sheets of the expansion for checking purposes. This, of course, has nothing to do with mathematics.

But in the last few years the Shanks figures have been rechecked, with the help of modern calculating machines, and an error has been found which affects the last couple of hundred figures (D. F. Ferguson, *Mathematical Gazette*, Vol. 30 (1946), p. 89). This new expansion has been added to the repertoire of the mathematician with the remarkable memory.

From the practical point of view we never need to know the value of π to more than a few decimal places. If we ever found ourselves stranded on a desert island, with an urgent need to know the expansion of π to *twenty* decimal places, the following rhyme would help:

> Sir, I bear a rhyme excelling
> In mystic force and magic spelling
> Celestial sprites elucidate
> All my own striving can't relate.

The number of letters in each word gives the corresponding digit in the expansion, with a comma for the decimal point, so that this mnemonic gives us

$$3.\ 14159\ 265358\ 979\ 323846,$$

which is more than enough.

Shanks's expansion did arouse some interest, before it was discovered to be incorrect, for the following reason. In the 608 figures which occur in the expansion, it might be expected that the nine digits and zero would each occur about the same number of times, that is about 61 times. But on counting, it was found that 3 occurs 68 times, 9 and 2 occur 67 times, 4 occurs 64 times, 1 and 6 occur 62 times, 0 occurs 60 times, 8 occurs 58 times, 5 occurs 56 times, and 7 occurs only 44 times.

It did seem, then, that 7 had something about it which the expansion did not care for. But nobody was bold enough to state that the expansion must be wrong, because the ordinary numbers are not treated equally in it. Yet it is a striking fact that this inequality in treatment is ironed out in the new and correct expansion! No law which gives the frequency of occurrence of any given digit in the expansion of π is known. It can be amusing to work out numbers to hundreds of decimal places. In Shanks's day it would have taken ten years of calculation to determine π to 1,000 decimal places. A short time ago four young men amused themselves over a weekend by calculating π to more than 2,000 places, with the help of an electronic computer. This was in the United States.

To return to circle-squarers. . . . . All mathematicians have been pestered at some time or other by people who have found solutions to problems which have not yet been solved, like the four-colour problem (Chapter V, p. 86). This kind of nuisance is not so unbearable as another, of far more frequent occurrence—being pestered by people who claim a large reward for proving something which has been disproved centuries ago! Augustus De Morgan, whom we have quoted with approval several times already, was a favourite target for circle-squarers. One in particular, an agricultural labourer, wrote to the Lord Chancellor, desiring his Lordship to hand over 100,000 pounds forthwith, the amount of the alleged reward for squaring the circle! The papers were sent to De Morgan, who wrote to the labourer, saying that he feared that the aspiring mathematician really did not know what he was talking about. The answer De Morgan received is worth quoting in full:—

"Doctor Morgan, Sir. Permit me to address you

Brute creation may perhaps enjoy the faculty of beholding visible things with a more penitrating eye than ourselves. But Spiritual objects are as far out of their reach as though they had no being.

Nearest therefore to the Brute creation are those men who Suppose themselves to be so far governed by external objects as to believe nothing but what they See and feel And Can accomedate to their Shallow understanding and Imaginations.

My Dear Sir Let us all Consult ourselves by the wise proverb.

I believe that every mans merit and ability aught to be appreciated and valued In proportion to its worth and utility

In whatever State or Circumstances they may fortunately or unfortunately be placed

And happy it is for evry man to know his worth and place

When a Gentleman of your Standing in Society Clad with those honours Can not understand or Solve a problem That is explicitly explained by words and Letters and mathematically operated by figuers He had best consult the wise proverb

Do that which thou Canst understand and Comprehend for thy good.

I would recommend that Such Gentleman Change his business And appropriate his time and attention to a Sunday School to Learn what he Could and keep the Little Children form durting their Close

With Sincere feelings of Gratitude for your weakness and Inability I am

Sir your Superior in Mathematics."

This letter was written in 1849. It has a Shavian flavour. We are more inhibited nowadays. Unlike the captain in the Royal Navy, writing against the Newtonian system of gravitation in 1833, we no longer call the Council of the Royal Society "craven dunghill cocks"! There is a high probability, of course, that the illustrious Council no longer deserves this picturesque appellation.

We end this chapter with some examples of the dangers which the inexperienced handler of infinite series may encounter. We begin with a series we have already encountered:

$$1 - 1 + 1 - 1 + 1 - 1 + \ldots\ldots\ldots$$

If we group the terms in one way, we have

$$S = (1 - 1) + (1 - 1) + \ldots\ldots$$
$$= \quad 0 \quad + \quad 0 \quad + \ldots\ldots$$
$$= \quad 0.$$

On the other hand, if we group the terms in a second way we have

$$S = 1 - (1 - 1) - (1 - 1) - \ldots\ldots$$
$$= 1 - \quad 0 \quad - \quad 0 \quad - \ldots\ldots$$
$$= 1.$$

By still another grouping

$$S = 1 - (1 - 1 + 1 - 1 + \ldots\ldots)$$
$$= 1 - \qquad\qquad S.$$

Therefore

$$2S = 1, \text{ or } S = \tfrac{1}{2}.$$

Here then is an infinite series whose sum is apparently any one of three quantities: 0, 1 or $\tfrac{1}{2}$. We saw, of course, that in reality we do not ascribe any sum to this series, but say it oscillates. Let us try another series. Let

$$S = 1 - 2 + 4 - 8 + 16 - 32 + 64 - 128 + \ldots\ldots$$

Then $S = 1 - 2 (1 - 2 + 4 - 8 + 16 - \ldots\ldots)$
$$= 1 - 2S.$$

Hence

$$3S = 1, \text{ or } S = \tfrac{1}{3}.$$

On the other hand, the series can be written

$$S = 1 + (-2 + 4) + (-8 + 16) + (-32 + 64) + \ldots.$$
$$= 1 + \quad 2 \quad + \quad 8 \quad + \quad 32 \quad + \ldots,$$

so that the sum can be made as large as we please, and greater than any assigned positive number. But we can also write

$$S = (1 - 2) + (4 - 8) + (16 - 32) + (64 - 128) + \ldots$$
$$= -1 \quad - \quad 4 \quad - \quad 8 \quad - \quad 64 \quad - \ldots,$$

and therefore S can be made less than any assigned negative number!

This series also oscillates, but whereas the first did so in a finite manner, this one oscillates *infinitely*. If we sum the first term, the first two terms, the first three terms, and so on, we obtain, in succession,

$$1, \; -1, \; 3, \; -5, \; 11, \; -21, \; \ldots \ldots$$

and it is evident that, as we go farther and farther on in the series, these *partial sums*, as they are called, jump from increasingly large positive numbers to increasingly small negative numbers. In a word, the series has no sum.

It is proper at this stage to say *precisely* what we mean by the *sum* of an infinite series. If we have an infinite series of terms

$$u_1 + u_2 + u_3 + u_4 + \ldots \ldots + u_n + \ldots \ldots$$

and form the partial sums

$$S_1 = u_1, \; S_2 = u_1 + u_2, \; S_3 = u_1 + u_2 + u_3, \ldots \ldots$$
$$S_n = u_1 + u_2 + \ldots \ldots \ldots + u_n,$$

then the numbers $S_1, S_2, S_3, \ldots \ldots, S_n$ are said to form a *sequence*. If, as n increases without limit (or, as we say, *tends to infinity*), S_n moves nearer and nearer to a definite number S, then we say that the infinite series *converges*, and that S is its sum.

To make the last part of this definition more precise, we say that the series is convergent and has the sum S if, given any positive number ϵ (epsilon) however small, there exists an integer n_o such that the numerical difference between S and S_n is less than ϵ as soon as $n > n_o$.

This definition serves to pin down the variable number S_n in a very practical way. We do not demand that it approach S and always be less than S, or always greater. It can be alternately less than S, or greater, with odd or even n. But it must remain as near S as we please for large enough n. For example, if the series is such that $S_n = (-1)^n/n$, then $S = 0$, and S_n is alternately positive and negative as n increases, but approaches steadily to zero.

It is not surprising, if we think about it, that an oscillating series which has no sum can be made to *appear* to have different sums. But we now consider a *convergent* series in which rearrangement of terms leads to curious results. The series is

$$1 - \tfrac{1}{2} + \tfrac{1}{3} - \tfrac{1}{4} + \tfrac{1}{5} - \tfrac{1}{6} + \ldots \ldots$$

The partial sum for an even number of terms, $2m$ say, is

$$S_{2m} = (1 - \tfrac{1}{2}) + (\tfrac{1}{3} - \tfrac{1}{4}) + \ldots + \left(\frac{1}{2m-1} - \frac{1}{2m} \right),$$

on bracketing the terms in pairs, and is therefore a sum of positive numbers, and is positive. On the other hand, we can also write

$$S_{2m} = 1 - (\tfrac{1}{2} - \tfrac{1}{3}) - (\tfrac{1}{4} - \tfrac{1}{5}) - \cdots - \frac{1}{2m},$$

which shows that S_{2m} is always less than 1. The first expression for S_{2m} shows that as m increases, so does this partial sum. Hence S_{2m} increases steadily with m, but is always less than 1, while the difference between S_{2m} and S_{2m+1} is $1/(2m+1)$, and this becomes infinitely small as m increases without limit, so that in deciding whether the series is convergent or not, we need only consider the partial sum S_{2m}.

Now, what can we say about a sequence of numbers which increase steadily, but are always less than a given fixed finite number, which is 1 in this case? Until the nineteenth century it was considered intuitively evident that such a sequence must be convergent: in other words, that there must exist a definite number S such that, as n increases without limit, S_n, the nth number of the sequence, approaches infinitely near to S. If we assume this theorem, it follows that the series we are considering above is convergent.

A proof of this fundamental theorem depends essentially on showing that S has a definite decimal expansion. The reader may wish to construct such a proof. We shall not give it here. We have now proved that the series

$$1 - \tfrac{1}{2} + \tfrac{1}{3} - \tfrac{1}{4} + \tfrac{1}{5} - \tfrac{1}{6} + \cdots$$

is convergent, and that its sum S is less than 1. Let us see what we can do by rearranging the terms! We have

$$2S = 2 - \tfrac{2}{2} + \tfrac{2}{3} - \tfrac{2}{4} + \tfrac{2}{5} - \tfrac{2}{6} + \tfrac{2}{7} - \tfrac{2}{8} + \tfrac{2}{9}$$
$$- \tfrac{2}{10} + \tfrac{2}{11} - \tfrac{2}{12} + \tfrac{2}{13} - \tfrac{2}{14} + \tfrac{2}{15} - \tfrac{2}{16} + \cdots$$
$$= 2 - 1 + \tfrac{2}{3} - \tfrac{1}{2} + \tfrac{2}{5} - \tfrac{1}{3} + \tfrac{2}{7} - \tfrac{1}{4} + \tfrac{2}{9}$$
$$- \tfrac{1}{5} + \tfrac{2}{11} - \tfrac{1}{6} + \tfrac{2}{13} - \tfrac{1}{7} + \tfrac{2}{15} - \tfrac{1}{8} + \cdots$$

before any rearrangement takes place. We now group terms with the same denominator. Then

$$2S = (2 - 1) - \tfrac{1}{2} + (\tfrac{2}{3} - \tfrac{1}{3}) - \tfrac{1}{4} + (\tfrac{2}{5} - \tfrac{1}{5}) - \tfrac{1}{6}$$
$$+ (\tfrac{2}{7} - \tfrac{1}{7}) - \tfrac{1}{8} + \cdots,$$

or $\qquad 2S = 1 - \tfrac{1}{2} + \tfrac{1}{3} - \tfrac{1}{4} + \tfrac{1}{5} - \tfrac{1}{6} + \tfrac{1}{7} - \tfrac{1}{8} + \cdots$

The series on the right is now the original series, and its sum is now $2S$ and not S! Moreover, if the operation of multiplying by 2 and collecting terms with the same denominator is repeated, the series can be summed to $4S$, $8S$, $16S$, ... !

Here then is a real dilemma—an infinite series which converges to a finite limit S less than 1, but which can be rearranged to converge to $2S$, $4S$, ... , and so on.

The difficulty arises from our attempt to apply to *infinite* series the processes of *finite* arithmetic. In finite arithmetic we assume that we may remove or insert brackets at will, grouping terms in any way we please. In other words we assume that

$$A + B + C = (A + B) + C = A + (B + C).$$

The contradictory results obtained above show that this regrouping cannot be applied to infinite series in general.

The question then arises—is it *ever* possible to rearrange and group the terms of a convergent infinite series with the certainty that its sum will not be changed? The answer is "Yes", provided that the series is *absolutely* convergent. An infinite series is absolutely convergent if the new series formed by *changing all the minus signs to plus signs converges*. It can then be shown that the original series itself converges, so that an absolutely convergent series converges! A series which consists of positive terms only is absolutely convergent if it is convergent.

Now the series we have been manipulating above is *not* absolutely convergent. If we change all the minus signs to plus signs, we obtain the series

$$1 + \tfrac{1}{2} + \tfrac{1}{3} + \tfrac{1}{4} + \tfrac{1}{5} + \ldots\ldots ,$$

and we saw earlier on that this harmonic series diverges, slowly but surely.

It was proved in 1854 by Riemann that the terms of a convergent but not absolutely convergent series can be so rearranged that the sum of the new series is any specified definite number, or so that the series diverges to positive or negative infinity! When we say that a series diverges to positive infinity, we mean that the partial sum S_n exceeds any given positive number however great, as soon as n exceeds an n_0 which can be determined. The series diverges to negative infinity if the same statement is true of $-S_n$.

The basis of Riemann's theorem is the fact that if a series converges, but not absolutely, the positive terms of the series, taken together, diverge to positive infinity, and the negative terms of the series, taken together, diverge to negative infinity. Hence by combining six of one with half a dozen of the other, we may arrive at any sum we please!

Having shown the pitfalls which beset the unwary manipulator of infinite series, it must also be remarked that safety precautions for handling these formidable creatures were only introduced after many a great mathematician had been severely mauled! It is, of course, possible to manipulate divergent series and to obtain correct mathematical results, and perhaps a too nervous approach to series is to be deprecated. But if one wishes to be *certain* that one's results are correct, it is advisable to keep to the rules which ensure safety.

We conclude this chapter by showing how the concept of an infinite geometric series resolves one of Zeno's paradoxes. In the fifth century B.C., Zeno of Elea enunciated a number of paradoxes, and the one we now consider has the title "Achilles and the Tortoise". We suppose that Achilles and the tortoise have a race, and that the tortoise is given a start of 100 yards. We also suppose that Achilles travels at the rate of 10 yards a second, and that the tortoise crawls at the rate of 1 yard a second. These rates are merely assumed for the purpose of illustrating the paradox.

Now Achilles covers the first 100 yards in 10 seconds dead; but in the meantime the tortoise has gone 10 yards. Achilles takes 1 second to cover this further distance, while the tortoise advances 1 yard. Achilles covers this distance in $\frac{1}{10}$ of a second, but the tortoise is still $\frac{1}{10}$ of a yard ahead! It takes Achilles, already a bit puffed, $\frac{1}{100}$ of a second to cover this $\frac{1}{10}$ of a yard, but the tortoise is still $\frac{1}{100}$ of a yard ahead, and so on. . . . . .

If we continue to argue in this way *verbally*, we never arrive at a point where Achilles overtakes the tortoise. Yet we know that he *does* overtake it! How do we resolve this paradox?

If we concentrate on finding *the time* Achilles takes to catch up with the tortoise, we have to sum the infinite geometric series

$$10 + 1 + \tfrac{1}{10} + \tfrac{1}{100} + \cdots \cdots,$$

and its sum to infinity is

$$\frac{10}{1 - \frac{1}{10}} = \frac{100}{9}.$$

We therefore know that after $11\frac{1}{9}$ seconds the tortoise has been overtaken. But it must be pointed out that this solution does not satisfy all philosophers, and many books on philosophy devote some space to a discussion of this Zeno paradox.

CHAPTER VIII

DOUBLE TALK

THE connection of mathematics with logic is notorious, and has been emphasised all through this book. In fact, one of the main points we have tried to make is that anyone who can argue logically can do mathematics. The converse of this statement—that anyone who is taught mathematics can argue logically—has been amply disproved. It is a sad fact that mathematicians will argue about politics, say, in a way that is a discredit to their training. A trivial example will be seized upon as a proof that the whole idea behind the National Health Service is wrong. Or, on the left wing, convincing arguments will be found to demonstrate that everything which is done by the Soviet government must be right. Educationists have long given up hope that the habits of thought which are fundamental in mathematics will be transferred to other spheres. Of course, this transfer often does take place. One of our finest judges is, or was a mathematician. But in general a training in mathematics will not achieve as much as a training in logic. We are now going to see whether logic itself can always be trusted!

In the sixth century B.C. Epimenides, the celebrated poet and prophet of Crete, is supposed to have made the remark: "All Cretans are liars." We see what we may deduce from this remark, first writing it in the form: "All statements made by Cretans are false."

Now Epimenides, who made this statement, was a Cretan. Hence, all statements made by him were false. In particular, his statement "All statements made by Cretans are false" was false, so that all statements made by Cretans were *not* false! What do we do now?

In order to see more clearly what is happening, it is better to express the argument thus:

(1) All statements made by Cretans are false;

(2) Statement (1) was made by a Cretan;

(3) Therefore statement (1) is false;

(4) Therefore all statements made by Cretans are not false.

Now statements (1) and (4) cannot both be true. Yet we have used only ordinary logical processes in deriving (4) from (1). Consequently statement (1) is self-contradictory.

The reader has probably heard of the village barber who declared that he shaved everyone in the village who did not shave himself, there being no other barber in the village. This statement seems to be unexceptionable until we ask: "Who shaves the barber?"

If he does not shave himself, then he is one of the people in the village who does not shave himself, and is therefore shaved by the barber, namely himself. If he shaves himself, he is naturally one of the people in the village who is not shaved by the barber!

Among modern paradoxes of this kind we have the following: Consider all the adjectives in the English language. Each has a certain meaning. In some adjectives the meaning applies to the adjective itself; in others it does not. Thus "short" is a short word, and "English" is an English word; but "French" is not a French word, and "hyphenated" is not a hyphenated word.

Since the meaning of an adjective must either apply to itself, or not, we can divide all adjectives into two classes. Let us call those which apply to themselves "autological", and those which do not apply to themselves "heterological". We now have two classes of adjectives, the autological and the heterological, and any adjective must belong to one or the other class.

Now what about the adjective "heterological"? If "heterological" is heterological, then this mere statement means that it applies to itself, and "heterological" must therefore be autological! On the other hand, if "heterological" is autological, then it does not apply to itself, and "heterological" must therefore be heterological!

More examples of this kind can be invented, but we have given proof that trouble can arise even with "pure logic", so much prized by those who never practise it, and we examine the common basis of the paradoxes we have described. The statements made above are all concerned with "all" the members of certain classes of things, and either the statements or the things to which the statements refer are themselves members of those classes.

To explain more precisely what we mean, we take the paradoxes in turn. In the first one we came across the statement "All statements made by Cretans are false." Since this is a statement made by a Cretan, it is a member of the class of all statements made by Cretans.

In the case of the village barber, we are concerned with the class of all men in the village who either shave themselves or do not shave themselves. The barber is evidently a member of this class. Finally, in the third paradox discussed above, the class of all adjectives, autological or heterological, obviously includes the adjective "heterological".

The vicious circle which is conjured into being as soon as a statement is made about "all" the members of a certain class, when the statement or the thing to which the statement refers is itself a member of the class, is difficult to avoid. As early as 1906 Bertrand Russell invented the theory of "logical types" in an attempt to underpin the foundations of logic. Russell pointed out that logical entities are not all of one type, but fall into a hierarchy of types, which are quite different, however similar they may appear to be. Moreover anything involving "all" of a certain class of objects is not of the same type as the objects themselves.

For example, take the case of "All statements made by Cretans are false." The statements referred to are statements about things. The statement *itself* is not a statement about things, but *a statement about statements of things*. It is therefore a statement of a different logical type, and we may therefore rule that it cannot refer to itself. This saves the given example from contradiction. But other examples of inherent difficulties in the theory of classes have turned up, and at one stage Russell attempted the construction of a logic which did not use the class-concept at all!

Russell's most famous paradox concerns the classes which are classes of themselves, or not. We can easily see that classes are either one or the other. For example, the class of all men is not a man, whereas the class of all ideas is an idea, and is therefore a member of the class of all ideas.

We can therefore divide all classes into two sets: those which are members of themselves being called S, and those which are not members of themselves being called T. The set T is therefore the class of all classes which are not members of themselves. Now T itself is a class. What can we say about it? It is either a member of itself, or it is not. If T *is* a member of itself it is a member of T, and is therefore one of the classes which are *not* members of themselves. On the other hand, if T is *not* a member of itself, it must be in the T class. But this statement itself shows that T *is* a member of itself. We know that any class is either a member of S or of T, and yet any statement we try to make about the classes S and T themselves becomes self-contradictory!

This Russell paradox has been relevant to the development of modern mathematics. The concept "class of all classes" had been very freely used before the Russell paradox exploded in the world of mathematics. One of the mathematicians apparently atomised by the explosion was Gottlob Frege, a German who had spent years in trying to put mathematics on a sound logical basis. His chief work was a two-volume treatise on the foundations of arithmetic, a treatise in which he used the notion of a class of all classes. Just as the second volume was about to appear, in 1903, Russell sent Frege the paradox we have just described. This was acknowledged at the end of the second volume, thus:

"A scientist can hardly meet with anything more undesirable than to have the foundation give way just as the work is finished. In this position I was put by a letter from Mr. Bertrand Russell just as the work was nearly through the press."

Frege's humility is outstanding in the history of mathematics, since mathematicians are not easily convinced that they are wrong, and are even more reluctant to acknowledge error in print. But it is pleasing to note that his work was not completely invalidated, and his reputation today stands very high indeed. Cantor was another mathematician whose work was apparently affected by the Russell and other paradoxes, but the magnificent superstructure he erected stands today unblemished, with no visible damage caused by the slight tremor which once shook it.

Russell is best known for his work on the foundations of mathematics. He made a valiant attempt to show that mathematics and logic are one, and that the whole of pure mathematics can be derived from logic. We now describe two concepts which are fundamental in mathematics, but which can hardly be said to belong to logic.

(1) *The principle of induction*: This may be stated as follows: "If a property is true for the number 1, and if it is established that it is true for $n + 1$ provided that it is true for n, then it will be true for all integers."

We used this principle in Chapter VII, p. 109, and it is the basis of many proofs in mathematics. The reader may wonder why it is called a *principle*, when it might appear that a theorem can be proved for any value of n by merely taking the successive steps necessary to get there. Of course we can do this for any *definite* value of n, but we cannot do this for *all n*, and so an assumption creeps in, which is stated in the principle.

(2) *Zermelo's axiom of choice*: In the early nineteen-hundreds Zermelo pointed out that many mathematical proofs depend on the ability to *choose* an element from a given set. For example on p. 114 we showed how the notion of the sum of an infinite geometric series is intuitive when we keep on bisecting a segment, and *choosing* a half to rub out. In this case there is no difficulty, since we can always choose the left-hand segment. But things are not always so simple, as we shall see. We first give the formal statement of the axiom:

Axiom of choice: If S is a collection of disjoint non-empty sets S_i, then there exists a set R which has as its elements exactly one element of each S_i.

Disjoint sets are merely sets with no elements in common. Suppose the set S consists of all the counties of England, and the elements S_i consist of the public-houses in these counties. Then it is possible to find a set R which has as its elements exactly *one* public-house from each county. We might *specify* the public-house to be the oldest in the county. Things are not so simple when we cannot, or do not, specify the element, and when the number of sets is infinite. For example, if we imagine a world in which there exists an infinite collection of pairs of shoes, does there exist a set R which contains just one shoe from each pair?

We may say that there does, since we can *define* R to consist of the *left* shoe of each pair. But suppose that this imaginary world also contains an infinite collection of pairs of socks! Does a corresponding set R exist for this collection? Unfortunately socks are manufactured in identical pairs, and it is impossible to distinguish a left sock from a right sock, so that we have no way of *defining* a set R which contains a representative of each pair of socks.

Despite this lack, we may still feel justified in asserting that such a set exists; but to do so is to employ the Axiom of Choice, whereas in the case of the shoes no such appeal was necessary.

In his introduction to the 1937 edition of "Principles of Mathematics" Bertrand Russell says:

"Whether this (the axiom of choice) is true or not, no one knows. It is easy to imagine universes in which it would be true, and it is impossible to prove that there are possible universes in which it would be false; but it is also impossible (at least, so I believe) to prove that there are no possible universes in which it would be false. I did not become aware of the necessity for this axiom until a year after the "Principles" was published. This book contains, in consequence, certain errors."

Poincaré, the great French mathematician, was scornful from the first of the attempt to derive mathematics from logic only. Poincaré was not only a great mathematician, but an amusing and scathing writer, if the necessity arose, and not afraid (unlike many mathematicians) to say what he thought. His celebrated "Science and Method" contains a devastating attack on those who think that the whole of mathematics may be derived from logic. The Nelson edition has a restrained introduction by Bertrand Russell.

We now give a brief description of some other important schools of thought on the foundations of mathematics. The school described above, the Frege-Russell school, is often called the School of Logisticism. There are two other Schools, of Intuitionism and of Formalism.

Intuitionism dates back to Kronecker (1823–1891), who insisted that mathematics is a construction on the basis of the "intuitively given" natural numbers. Kronecker's endeavour to force everything mathematical into the pattern of number theory is illustrated by his well-known statement at a meeting in Berlin in 1886: "Die ganzen Zahlen hat der liebe Gott gemacht, alles andere ist Menchenwerk."*

The greatest present-day exponent of this point of view is the Dutch mathematician Brouwer. The most striking aspect of Intuitionism is what might be called its *self-sufficiency*. It relies in no way on other philosophies or logic. Its basic ideas are to be found in the *intuition*, which seems to be similar to the time (not the spatial)-intuition of Kant. Specifically it recognises the ability of the individual person to perform a series of mental acts consisting of a first act, then another, then another, and so on. In this way one attains "fundamental series", the best known of which is the series of natural numbers.

This operation is not dependent on the use of a language. For the *communication* of mathematics the usual symbolic devices, including ordinary language, are necessary, but this is their only function. This attitude seems to make mathematics virtually an individual affair rather than an organised or cultural phenomenon. Or possibly the emphasis is on the freeing of the foundations from linguistic influence.

One of the most interesting, and startling, aspects of Intuitionist logic is the general refusal to accept the law of the excluded middle, and its consequence, which we used freely in Chapter IV, and in other chapters, that the negation of the negation of a proposition p implies p. In other words, a double negative gives a positive, or in symbols:

$$(p')' = p.$$

This law is universally used in ordinary mathematics. A method of proving a theorem is to begin by assuming the theorem is false. If it is possible to prove that the assumption of the falsity of the theorem leads to contradiction, then the theorem is true. For any proposition which leads to contradiction must be false. Hence if the falsity of the theorem is false, then the theorem is true by the principle cited above

---

* "The integers have been made by God. All else is the work of man."

(the negation of the negation of p implies p). All proofs by *reductio ad absurdum* are of this type.

The intuitionist does not accept this kind of proof. He accepts the principle that any proposition which implies contradiction must be false, but not the Law of the Excluded Middle which says that $(p')' = p$. Hence much in modern mathematics is rejected by the Brouwer school, but a remarkable number of theorems have been proved by their methods. Those portions of mathematical analysis which are constructible by actual computational methods are, in general, to be found in intuitionist mathematics.

It is so generally accepted that a proposition is either true or false, which is what the Law of the Excluded Middle implies, that it is as well to give an example of a theorem on which we cannot give any verdict:

A number N is defined as follows: "N is a number such that $0 \leqslant N \leqslant 1$. In decimal form its nth digit after the decimal point, a_n, is 0, unless the nth digit of the decimal part of π is the first of a sequence of seven sevens — 7777777 — in which case the digit a_n is 1."

If N is defined in this way, we cannot say whether $N = 0$ or not, since we have no way of knowing that there does not exist a sequence of seven sevens somewhere in the decimal expansion of π. Hence we cannot prove that $N = 0$, nor can we prove that $N \neq 0$. Of course, at some future time one of the two theorems may be provable.

In contrast to the intuitionist tenet that language and symbolism are not basic to mathematics stands the formalist conviction that symbols, and operations with them, constitute the very heart of the matter. To explain the formalist view, we must go back to Hilbert's early work on the axiomatisation of mathematics.

It is possible to choose a set of axioms for geometry, and to develop Euclidean geometry, say, on the basis of these axioms. An axiom is a proposition which we *accept*, and do not attempt to prove. Now we want our axioms to be independent, so that one axiom cannot be derived from the others, and it is also essential that some of the axioms, taken by themselves, should not indicate a result which contradicts another of the axioms! In other words, we ask that the mathematics built up from a set of axioms should be free from contradiction. For example, if our mathematics dealt with certain undefined entities called *points* and *lines*, and one of our axioms asserted that a line contained not more than two points, we should not be happy if we could prove, from the other axioms, the existence of a line containing three points.

The first requirement, that the axioms be independent of each other, is a practical and aesthetic one. Utility and beauty are closely linked in mathematics. To show that a given set of axioms are free from contradiction is not usually an easy problem. One method of proof is the discovery of an existing mathematics which satisfies the basic axioms. If this mathematics is free from contradiction, then so are the basic axioms. Such a mathematics is called a *model* for the set of axioms. But the proof that known mathematical models, such as real number arithmetic, are free from contradiction is far from easy, and in the case of the given example this proof has not yet been given! So that all difficulties cannot be overcome by developing mathematics on an axiomatic basis.

To cope with these difficulties Hilbert decided on a union of the axiomatic and logistic methods, and built up a subject called *metamathematics*. This deals with the *proofs* of ordinary mathematics, these proofs themselves forming the subject of investigation. In any mathematical theory only certain methods of proof are to be allowed, and if it can be shown that no formula (in a certain technical sense) *and* the negation of this formula can arise from these methods of proof, then the mathematical system can be said to be consistent.

For certain elementary mathematical systems the proofs of consistency have been successfully carried out. For instance, it has been proved that a certain elementary system of axioms involving integers is consistent. But this has not yet been extended to the entire arithmetic of integers. In any case, whether the complete Hilbert programme can be carried out has been made very doubtful by the work of Goedel. His results, published in 1931, may be roughly characterised as a demonstration that, in any system broad enough to contain all the formulas of a formalised elementary number theory, there exist theorems (formulas) which can neither be proved nor disproved within the system.

We have probably said enough to show that the foundations of mathematics is a vital and developing subject. Although it has not yet been proved that the mathematics they do is free from contradiction, most mathematicians pursue their researches quite happily, unworried by any fear that one day the whole magnificent cultural superstructure will topple on them. It is even probably true to say of the worker on the foundations of mathematics that he is more interested in the problems to which they have given rise than in the choice or validity of an underlying philosophy.

CHAPTER IX

WHAT IS MATHEMATICS?

ENOUGH has been said in this book to show that, whatever mathematics may be, it is an international activity. This has not always been the case. The connection of the Greeks with geometry is well-known. Euclid, in a modified form, is still taught in schools. The Romans, although influenced by Greek mathematics, had their own system of writing numerals, a system much more cumbersome than the Greek for numerical manipulation, yet surviving to this day on monuments and title pages.

In the Chinese culture no geometry of the Greek type was known. Mathematics in China consisted, in the main, of numerical computations and the solution of algebraic equations. Although there were evidently contacts between Eastern and Western civilisations from early Christian times onwards, very few mathematical ideas passed between the two cultures until comparatively recent times. The Greeks, as we said above, developed geometry. On the other hand, no systematic development of algebra took place in the Mediterranean area during the period in which Greek culture flourished, nor for centuries afterwards. It was the Arabs who preserved and transmitted, via Africa and Spain, the Hindu-Arabic mathematics, and the system of numeration which is universal today, together with Greek geometry.

What is mathematics? Let us begin our answer to this question by looking at what mathematicians do. If we examine any copy of "Mathematical Reviews", a publication which exists solely for the purpose of giving short reviews of mathematical books or papers, we see that several hundred papers or books are reviewed every month. The Table of Contents is very expressive of the manifold activities of present-day mathematicians. We reproduce it, but not exactly as it is printed. The subject-headings are:

History: Foundations: Algebra: Number Theory: Analysis: Topology: Geometry: Numerical and Graphical Methods: Relativity: Mechanics: Mathematical Physics.

Under Algebra there are two subtitles: Abstract Algebra, and Theory of Groups. Under Analysis there are sixteen subheadings, beginning with Calculus and ending with Mathematical Biology.

Geometry has three subheadings, Mechanics two, and Mathematical Physics has two.

The books and papers reviewed in Mathematical Reviews are in English, French, Italian, German, Dutch, Polish, Russian, Chinese, Japanese, Hebrew, . . . and in fact in every language in which scientific journals are published. Reviews are published in English, French, German and Italian, and every research mathematician has to be familiar with these languages *as written*. This is not as difficult as it may seem, because most mathematical papers employ only a special, limited vocabulary, and a familiarity with a branch of mathematics enables a mathematician to read a paper in that branch, or to have a good guess at its contents, in practically any tongue.

It is therefore clear that mathematicians from all parts of the world can understand each other, if they are working on the same kind of mathematics. International Conferences, at which personal contacts can be made, are held every four years. The last two were held in Cambridge, Mass., U.S.A. in 1950, and in Amsterdam in 1954. The attendance at each of these Conferences was well over 2,000. Two prizes of great value are awarded at each Conference, for the best work done by a mathematician under forty. At the Amsterdam Conference these prizes went to a Japanese and a Frenchman. The jury always consists of a selection of the world's greatest mathematicians. Their task is not an easy one, nor is their decision universally acclaimed on all occasions. Each Conference has its *prima donna*, and the human aspect of these large gatherings of mathematical specialists is one of the great attractions. To see *A*, who has always bitterly criticised *B*, beaming at one of *B*'s lectures because *B* has referred to *A*'s work as "classical"; to hear the great *C*, whose day is, alas, no more, gently and wittily reminding his audience that all the present-day work being done in a certain field was anticipated by him in 1905, and that it is polite to acknowledge one's sources—all this is great fun.

What kind of human beings become mathematicians? In some cases it is probably true that a man turns to mathematics as a compensation for an inability to face the external world. But there are mathematicians of all sorts—mathematicians who are genial giants, mathematicians who look like rotund Rotarians, mathematicians who are tone-deaf, mathematicians who adore Bach, mathematicians who play jazz, mathematicians who dance the samba, mathematicians who are obviously mathematicians, and nothing else, and mathematicians you would never suspect of being mathematicians. The

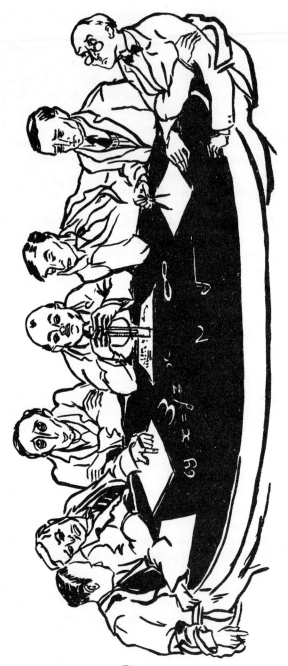

Fig. 42.

one faculty common to all mathematicians is a capacity for abstract thought.

We have still not answered the question "What is mathematics?" But we can certainly say what it is not! The logical positivists have had their say about mathematics. This school of philosophers only assigns meaning to statements which can be tested in some definite way, and have thus disposed of much which has troubled great minds for thousands of years. Unfortunately, they can give no reason why their criterion is better than any other, nor do they indicate how *it* can be tested. But if we accept their hypotheses, and see what they have to offer about the nature of mathematics, we must be disappointed. An infinitely intelligent person, contends Professor Ayer,* would find mathematics dull, since he would be able to see at a glance all the possible consequences of any set of axioms!

Speaking as a logical positivist, this is a meaningless statement, since we cannot devise any test to see whether any person is infinitely intelligent. But if we allow meaning to the statement, it assumes that mathematical development is exclusively occupied with logical deductions from given axioms or theorems. Let us see what Poincaré says about this in his "Science and Method".

"What, in fact, is mathematical discovery? It does not consist in making new combinations with mathematical entities that are already known. That can be done by anyone, and the combinations that could be so formed would be infinite in number, and the greater part of them would be absolutely devoid of interest. Discovery consists precisely in not constructing useless combinations, but in constructing those which are useful, which are an infinitely small minority. Discovery is discernment, selection.

"Mathematical facts worthy of being studied are those which, by their analogy with other facts, are capable of leading us to the knowledge of a mathematical law, in the same way that experimental facts lead us to the knowledge of a physical law. They are those which reveal unsuspected relations between other facts, long since known, but wrongly believed to be unrelated to each other."

No mathematician would wish to add anything to this statement.

* "Language, Truth and Logic", 1947 edn., p. 85.

INDEX

A CATALOG OF SELECTED
DOVER BOOKS
IN ALL FIELDS OF INTEREST

A CATALOG OF SELECTED DOVER
BOOKS IN ALL FIELDS OF INTEREST

DRAWINGS OF REMBRANDT, edited by Seymour Slive. Updated Lippmann, Hofstede de Groot edition, with definitive scholarly apparatus. All portraits, biblical sketches, landscapes, nudes. Oriental figures, classical studies, together with selection of work by followers. 550 illustrations. Total of 630pp. 9⅛ × 12¼.
21485-0, 21486-9 Pa., Two-vol. set $25.00

GHOST AND HORROR STORIES OF AMBROSE BIERCE, Ambrose Bierce. 24 tales vividly imagined, strangely prophetic, and decades ahead of their time in technical skill: "The Damned Thing," "An Inhabitant of Carcosa," "The Eyes of the Panther," "Moxon's Master," and 20 more. 199pp. 5⅜ × 8½. 20767-6 Pa. $3.95

ETHICAL WRITINGS OF MAIMONIDES, Maimonides. Most significant ethical works of great medieval sage, newly translated for utmost precision, readability. Laws Concerning Character Traits, Eight Chapters, more. 192pp. 5⅜ × 8½.
24522-5 Pa. $4.50

THE EXPLORATION OF THE COLORADO RIVER AND ITS CANYONS, J. W. Powell. Full text of Powell's 1,000-mile expedition down the fabled Colorado in 1869. Superb account of terrain, geology, vegetation, Indians, famine, mutiny, treacherous rapids, mighty canyons, during exploration of last unknown part of continental U.S. 400pp. 5⅜ × 8½. 20094-9 Pa. $6.95

HISTORY OF PHILOSOPHY, Julián Marías. Clearest one-volume history on the market. Every major philosopher and dozens of others, to Existentialism and later. 505pp. 5⅜ × 8½. 21739-6 Pa. $8.50

ALL ABOUT LIGHTNING, Martin A. Uman. Highly readable non-technical survey of nature and causes of lightning, thunderstorms, ball lightning, St. Elmo's Fire, much more. Illustrated. 192pp. 5⅜ × 8½. 25237-X Pa. $5.95

SAILING ALONE AROUND THE WORLD, Captain Joshua Slocum. First man to sail around the world, alone, in small boat. One of great feats of seamanship told in delightful manner. 67 illustrations. 294pp. 5⅜ × 8½. 20326-3 Pa. $4.50

LETTERS AND NOTES ON THE MANNERS, CUSTOMS AND CONDITIONS OF THE NORTH AMERICAN INDIANS, George Catlin. Classic account of life among Plains Indians: ceremonies, hunt, warfare, etc. 312 plates. 572pp. of text. 6⅛ × 9¼. 22118-0, 22119-9 Pa. Two-vol. set $15.90

ALASKA: The Harriman Expedition, 1899, John Burroughs, John Muir, et al. Informative, engrossing accounts of two-month, 9,000-mile expedition. Native peoples, wildlife, forests, geography, salmon industry, glaciers, more. Profusely illustrated. 240 black-and-white line drawings. 124 black-and-white photographs. 3 maps. Index. 576pp. 5⅜ × 8½. 25109-8 Pa. $11.95

THE BOOK OF BEASTS: Being a Translation from a Latin Bestiary of the Twelfth Century, T. H. White. Wonderful catalog real and fanciful beasts: manticore, griffin, phoenix, amphivius, jaculus, many more. White's witty erudite commentary on scientific, historical aspects. Fascinating glimpse of medieval mind. Illustrated. 296pp. 5⅜ × 8¼. (Available in U.S. only)   24609-4 Pa. $5.95

FRANK LLOYD WRIGHT: ARCHITECTURE AND NATURE With 160 Illustrations, Donald Hoffmann. Profusely illustrated study of influence of nature—especially prairie—on Wright's designs for Fallingwater, Robie House, Guggenheim Museum, other masterpieces. 96pp. 9¼ × 10¾.   25098-9 Pa. $7.95

FRANK LLOYD WRIGHT'S FALLINGWATER, Donald Hoffmann. Wright's famous waterfall house: planning and construction of organic idea. History of site, owners, Wright's personal involvement. Photographs of various stages of building. Preface by Edgar Kaufmann, Jr. 100 illustrations. 112pp. 9¼ × 10.
23671-4 Pa. $7.95

YEARS WITH FRANK LLOYD WRIGHT: Apprentice to Genius, Edgar Tafel. Insightful memoir by a former apprentice presents a revealing portrait of Wright the man, the inspired teacher, the greatest American architect. 372 black-and-white illustrations. Preface. Index. vi + 228pp. 8¼ × 11.   24801-1 Pa. $9.95

THE STORY OF KING ARTHUR AND HIS KNIGHTS, Howard Pyle. Enchanting version of King Arthur fable has delighted generations with imaginative narratives of exciting adventures and unforgettable illustrations by the author. 41 illustrations. xviii + 313pp. 6⅛ × 9¼.   21445-1 Pa. $5.95

THE GODS OF THE EGYPTIANS, E. A. Wallis Budge. Thorough coverage of numerous gods of ancient Egypt by foremost Egyptologist. Information on evolution of cults, rites and gods; the cult of Osiris; the Book of the Dead and its rites; the sacred animals and birds; Heaven and Hell; and more. 956pp. 6⅛ × 9¼.
22055-9, 22056-7 Pa., Two-vol. set $20.00

A THEOLOGICO-POLITICAL TREATISE, Benedict Spinoza. Also contains unfinished *Political Treatise*. Great classic on religious liberty, theory of government on common consent. R. Elwes translation. Total of 421pp. 5⅜ × 8½.
20249-6 Pa. $6.95

INCIDENTS OF TRAVEL IN CENTRAL AMERICA, CHIAPAS, AND YUCATAN, John L. Stephens. Almost single-handed discovery of Maya culture; exploration of ruined cities, monuments, temples; customs of Indians. 115 drawings. 892pp. 5⅜ × 8½.   22404-X, 22405-8 Pa., Two-vol. set $15.90

LOS CAPRICHOS, Francisco Goya. 80 plates of wild, grotesque monsters and caricatures. Prado manuscript included. 183pp. 6⅞ × 9⅜.   22384-1 Pa. $4.95

AUTOBIOGRAPHY: The Story of My Experiments with Truth, Mohandas K. Gandhi. Not hagiography, but Gandhi in his own words. Boyhood, legal studies, purification, the growth of the Satyagraha (nonviolent protest) movement. Critical, inspiring work of the man who freed India. 480pp. 5⅜ × 8½. (Available in U.S. only)
24593-4 Pa. $6.95

ILLUSTRATED DICTIONARY OF HISTORIC ARCHITECTURE, edited by Cyril M. Harris. Extraordinary compendium of clear, concise definitions for over 5,000 important architectural terms complemented by over 2,000 line drawings. Covers full spectrum of architecture from ancient ruins to 20th-century Modernism. Preface. 592pp. 7½ × 9⅝. 24444-X Pa. $14.95

THE NIGHT BEFORE CHRISTMAS, Clement Moore. Full text, and woodcuts from original 1848 book. Also critical, historical material. 19 illustrations. 40pp. 4⅝ × 6. 22797-9 Pa. $2.25

THE LESSON OF JAPANESE ARCHITECTURE: 165 Photographs, Jiro Harada. Memorable gallery of 165 photographs taken in the 1930's of exquisite Japanese homes of the well-to-do and historic buildings. 13 line diagrams. 192pp. 8⅜ × 11¼. 24778-3 Pa. $8.95

THE AUTOBIOGRAPHY OF CHARLES DARWIN AND SELECTED LETTERS, edited by Francis Darwin. The fascinating life of eccentric genius composed of an intimate memoir by Darwin (intended for his children); commentary by his son, Francis; hundreds of fragments from notebooks, journals, papers; and letters to and from Lyell, Hooker, Huxley, Wallace and Henslow. xi + 365pp. 5⅜ × 8. 20479-0 Pa. $5.95

WONDERS OF THE SKY: Observing Rainbows, Comets, Eclipses, the Stars and Other Phenomena, Fred Schaaf. Charming, easy-to-read poetic guide to all manner of celestial events visible to the naked eye. Mock suns, glories, Belt of Venus, more. Illustrated. 299pp. 5¼ × 8¼. 24402-4 Pa. $7.95

BURNHAM'S CELESTIAL HANDBOOK, Robert Burnham, Jr. Thorough guide to the stars beyond our solar system. Exhaustive treatment. Alphabetical by constellation: Andromeda to Cetus in Vol. 1; Chamaeleon to Orion in Vol. 2; and Pavo to Vulpecula in Vol. 3. Hundreds of illustrations. Index in Vol. 3. 2,000pp. 6⅛ × 9¼. 23567-X, 23568-8, 23673-0 Pa., Three-vol. set $36.85

STAR NAMES: Their Lore and Meaning, Richard Hinckley Allen. Fascinating history of names various cultures have given to constellations and literary and folkloristic uses that have been made of stars. Indexes to subjects. Arabic and Greek names. Biblical references. Bibliography. 563pp. 5⅜ × 8½. 21079-0 Pa. $7.95

THIRTY YEARS THAT SHOOK PHYSICS: The Story of Quantum Theory, George Gamow. Lucid, accessible introduction to influential theory of energy and matter. Careful explanations of Dirac's anti-particles, Bohr's model of the atom, much more. 12 plates. Numerous drawings. 240pp. 5⅜ × 8½. 24895-X Pa. $4.95

CHINESE DOMESTIC FURNITURE IN PHOTOGRAPHS AND MEASURED DRAWINGS, Gustav Ecke. A rare volume, now affordably priced for antique collectors, furniture buffs and art historians. Detailed review of styles ranging from early Shang to late Ming. Unabridged republication. 161 black-and-white drawings, photos. Total of 224pp. 8⅜ × 11¼. (Available in U.S. only) 25171-3 Pa. $12.95

VINCENT VAN GOGH: A Biography, Julius Meier-Graefe. Dynamic, penetrating study of artist's life, relationship with brother, Theo, painting techniques, travels, more. Readable, engrossing. 160pp. 5⅜ × 8½. (Available in U.S. only) 25253-1 Pa. $3.95

HOW TO WRITE, Gertrude Stein. Gertrude Stein claimed anyone could understand her unconventional writing—here are clues to help. Fascinating improvisations, language experiments, explanations illuminate Stein's craft and the art of writing. Total of 414pp. 4⅝ × 6⅜. 23144-5 Pa. $5.95

ADVENTURES AT SEA IN THE GREAT AGE OF SAIL: Five Firsthand Narratives, edited by Elliot Snow. Rare true accounts of exploration, whaling, shipwreck, fierce natives, trade, shipboard life, more. 33 illustrations. Introduction. 353pp. 5⅜ × 8½. 25177-2 Pa. $7.95

THE HERBAL OR GENERAL HISTORY OF PLANTS, John Gerard. Classic descriptions of about 2,850 plants—with over 2,700 illustrations—includes Latin and English names, physical descriptions, varieties, time and place of growth, more. 2,706 illustrations. xlv + 1,678pp. 8½ × 12¼. 23147-X Cloth. $75.00

DOROTHY AND THE WIZARD IN OZ, L. Frank Baum. Dorothy and the Wizard visit the center of the Earth, where people are vegetables, glass houses grow and Oz characters reappear. Classic sequel to *Wizard of Oz*. 256pp. 5⅜ × 8.
24714-7 Pa. $4.95

SONGS OF EXPERIENCE: Facsimile Reproduction with 26 Plates in Full Color, William Blake. This facsimile of Blake's original "Illuminated Book" reproduces 26 full-color plates from a rare 1826 edition. Includes "The Tyger," "London," "Holy Thursday," and other immortal poems. 26 color plates. Printed text of poems. 48pp. 5¼ × 7. 24636-1 Pa. $3.50

SONGS OF INNOCENCE, William Blake. The first and most popular of Blake's famous "Illuminated Books," in a facsimile edition reproducing all 31 brightly colored plates. Additional printed text of each poem. 64pp. 5¼ × 7.
22764-2 Pa. $3.50

PRECIOUS STONES, Max Bauer. Classic, thorough study of diamonds, rubies, emeralds, garnets, etc.: physical character, occurrence, properties, use, similar topics. 20 plates, 8 in color. 94 figures. 659pp. 6⅛ × 9¼.
21910-0, 21911-9 Pa., Two-vol. set $14.90

ENCYCLOPEDIA OF VICTORIAN NEEDLEWORK, S. F. A. Caulfeild and Blanche Saward. Full, precise descriptions of stitches, techniques for dozens of needlecrafts—most exhaustive reference of its kind. Over 800 figures. Total of 679pp. 8⅛ × 11. Two volumes. Vol. 1 22800-2 Pa. $10.95
Vol. 2 22801-0 Pa. $10.95

THE MARVELOUS LAND OF OZ, L. Frank Baum. Second Oz book, the Scarecrow and Tin Woodman are back with hero named Tip, Oz magic. 136 illustrations. 287pp. 5⅜ × 8½. 20692-0 Pa. $5.95

WILD FOWL DECOYS, Joel Barber. Basic book on the subject, by foremost authority and collector. Reveals history of decoy making and rigging, place in American culture, different kinds of decoys, how to make them, and how to use them. 140 plates. 156pp. 7⅞ × 10¾. 20011-6 Pa. $7.95

HISTORY OF LACE, Mrs. Bury Palliser. Definitive, profusely illustrated chronicle of lace from earliest times to late 19th century. Laces of Italy, Greece, England, France, Belgium, etc. Landmark of needlework scholarship. 266 illustrations. 672pp. 6⅛ × 9¼. 24742-2 Pa. $14.95

ILLUSTRATED GUIDE TO SHAKER FURNITURE, Robert Meader. All furniture and appurtenances, with much on unknown local styles. 235 photos. 146pp. 9 × 12.   22819-3 Pa. $7.95

WHALE SHIPS AND WHALING: A Pictorial Survey, George Francis Dow. Over 200 vintage engravings, drawings, photographs of barks, brigs, cutters, other vessels. Also harpoons, lances, whaling guns, many other artifacts. Comprehensive text by foremost authority. 207 black-and-white illustrations. 288pp. 6 × 9.
24808-9 Pa. $8.95

THE BERTRAMS, Anthony Trollope. Powerful portrayal of blind self-will and thwarted ambition includes one of Trollope's most heartrending love stories. 497pp. 5⅜ × 8½.   25119-5 Pa. $8.95

ADVENTURES WITH A HAND LENS, Richard Headstrom. Clearly written guide to observing and studying flowers and grasses, fish scales, moth and insect wings, egg cases, buds, feathers, seeds, leaf scars, moss, molds, ferns, common crystals, etc.—all with an ordinary, inexpensive magnifying glass. 209 exact line drawings aid in your discoveries. 220pp. 5⅜ × 8½.   23330-8 Pa. $3.95

RODIN ON ART AND ARTISTS, Auguste Rodin. Great sculptor's candid, wide-ranging comments on meaning of art; great artists; relation of sculpture to poetry, painting, music; philosophy of life, more. 76 superb black-and-white illustrations of Rodin's sculpture, drawings and prints. 119pp. 8⅜ × 11¼.   24487-3 Pa. $6.95

FIFTY CLASSIC FRENCH FILMS, 1912–1982: A Pictorial Record, Anthony Slide. Memorable stills from Grand Illusion, Beauty and the Beast, Hiroshima, Mon Amour, many more. Credits, plot synopses, reviews, etc. 160pp. 8¼ × 11.
25256-6 Pa. $11.95

THE PRINCIPLES OF PSYCHOLOGY, William James. Famous long course complete, unabridged. Stream of thought, time perception, memory, experimental methods; great work decades ahead of its time. 94 figures. 1,391pp. 5⅜ × 8½.
20381-6, 20382-4 Pa., Two-vol. set $19.90

BODIES IN A BOOKSHOP, R. T. Campbell. Challenging mystery of blackmail and murder with ingenious plot and superbly drawn characters. In the best tradition of British suspense fiction. 192pp. 5⅜ × 8½.   24720-1 Pa. $3.95

CALLAS: PORTRAIT OF A PRIMA DONNA, George Jellinek. Renowned commentator on the musical scene chronicles incredible career and life of the most controversial, fascinating, influential operatic personality of our time. 64 black-and-white photographs. 416pp. 5⅜ × 8¼.   25047-4 Pa. $7.95

GEOMETRY, RELATIVITY AND THE FOURTH DIMENSION, Rudolph Rucker. Exposition of fourth dimension, concepts of relativity as Flatland characters continue adventures. Popular, easily followed yet accurate, profound. 141 illustrations. 133pp. 5⅜ × 8½.   23400-2 Pa. $3.50

HOUSEHOLD STORIES BY THE BROTHERS GRIMM, with pictures by Walter Crane. 53 classic stories—Rumpelstiltskin, Rapunzel, Hansel and Gretel, the Fisherman and his Wife, Snow White, Tom Thumb, Sleeping Beauty, Cinderella, and so much more—lavishly illustrated with original 19th century drawings. 114 illustrations. x + 269pp. 5⅜ × 8½.   21080-4 Pa. $4.50

SUNDIALS, Albert Waugh. Far and away the best, most thorough coverage of ideas, mathematics concerned, types, construction, adjusting anywhere. Over 100 illustrations. 230pp. 5⅜ × 8½. 22947-5 Pa. $4.00

PICTURE HISTORY OF THE NORMANDIE: With 190 Illustrations, Frank O. Braynard. Full story of legendary French ocean liner: Art Deco interiors, design innovations, furnishings, celebrities, maiden voyage, tragic fire, much more. Extensive text. 144pp. 8⅜ × 11¾. 25257-4 Pa. $9.95

THE FIRST AMERICAN COOKBOOK: A Facsimile of "American Cookery," 1796, Amelia Simmons. Facsimile of the first American-written cookbook published in the United States contains authentic recipes for colonial favorites—pumpkin pudding, winter squash pudding, spruce beer, Indian slapjacks, and more. Introductory Essay and Glossary of colonial cooking terms. 80pp. 5⅜ × 8½. 24710-4 Pa. $3.50

101 PUZZLES IN THOUGHT AND LOGIC, C. R. Wylie, Jr. Solve murders and robberies, find out which fishermen are liars, how a blind man could possibly identify a color—purely by your own reasoning! 107pp. 5⅜ × 8½. 20367-0 Pa. $2.00

THE BOOK OF WORLD-FAMOUS MUSIC—CLASSICAL, POPULAR AND FOLK, James J. Fuld. Revised and enlarged republication of landmark work in musico-bibliography. Full information about nearly 1,000 songs and compositions including first lines of music and lyrics. New supplement. Index. 800pp. 5⅜ × 8¼. 24857-7 Pa. $14.95

ANTHROPOLOGY AND MODERN LIFE, Franz Boas. Great anthropologist's classic treatise on race and culture. Introduction by Ruth Bunzel. Only inexpensive paperback edition. 255pp. 5⅜ × 8½. 25245-0 Pa. $5.95

THE TALE OF PETER RABBIT, Beatrix Potter. The inimitable Peter's terrifying adventure in Mr. McGregor's garden, with all 27 wonderful, full-color Potter illustrations. 55pp. 4¼ × 5½. (Available in U.S. only) 22827-4 Pa. $1.75

THREE PROPHETIC SCIENCE FICTION NOVELS, H. G. Wells. *When the Sleeper Wakes, A Story of the Days to Come* and *The Time Machine* (full version). 335pp. 5⅜ × 8½. (Available in U.S. only) 20605-X Pa. $5.95

APICIUS COOKERY AND DINING IN IMPERIAL ROME, edited and translated by Joseph Dommers Vehling. Oldest known cookbook in existence offers readers a clear picture of what foods Romans ate, how they prepared them, etc. 49 illustrations. 301pp. 6⅛ × 9¼. 23563-7 Pa. $6.00

SHAKESPEARE LEXICON AND QUOTATION DICTIONARY, Alexander Schmidt. Full definitions, locations, shades of meaning of every word in plays and poems. More than 50,000 exact quotations. 1,485pp. 6½ × 9¼. 22726-X, 22727-8 Pa., Two-vol. set $27.90

THE WORLD'S GREAT SPEECHES, edited by Lewis Copeland and Lawrence W. Lamm. Vast collection of 278 speeches from Greeks to 1970. Powerful and effective models; unique look at history. 842pp. 5⅜ × 8½. 20468-5 Pa. $10.95

THE BLUE FAIRY BOOK, Andrew Lang. The first, most famous collection, with many familiar tales: Little Red Riding Hood, Aladdin and the Wonderful Lamp, Puss in Boots, Sleeping Beauty, Hansel and Gretel, Rumpelstiltskin; 37 in all. 138 illustrations. 390pp. 5⅜ × 8½.　21437-0 Pa. $5.95

THE STORY OF THE CHAMPIONS OF THE ROUND TABLE, Howard Pyle. Sir Launcelot, Sir Tristram and Sir Percival in spirited adventures of love and triumph retold in Pyle's inimitable style. 50 drawings, 31 full-page. xviii + 329pp. 6½ × 9¼.　21883-X Pa. $6.95

AUDUBON AND HIS JOURNALS, Maria Audubon. Unmatched two-volume portrait of the great artist, naturalist and author contains his journals, an excellent biography by his granddaughter, expert annotations by the noted ornithologist, Dr. Elliott Coues, and 37 superb illustrations. Total of 1,200pp. 5⅜ × 8.
Vol. I 25143-8 Pa. $8.95
Vol. II 25144-6 Pa. $8.95

GREAT DINOSAUR HUNTERS AND THEIR DISCOVERIES, Edwin H. Colbert. Fascinating, lavishly illustrated chronicle of dinosaur research, 1820's to 1960. Achievements of Cope, Marsh, Brown, Buckland, Mantell, Huxley, many others. 384pp. 5¼ × 8¼.　24701-5 Pa. $6.95

THE TASTEMAKERS, Russell Lynes. Informal, illustrated social history of American taste 1850's–1950's. First popularized categories Highbrow, Lowbrow, Middlebrow. 129 illustrations. New (1979) afterword. 384pp. 6 × 9.
23993-4 Pa. $6.95

DOUBLE CROSS PURPOSES, Ronald A. Knox. A treasure hunt in the Scottish Highlands, an old map, unidentified corpse, surprise discoveries keep reader guessing in this cleverly intricate tale of financial skullduggery. 2 black-and-white maps. 320pp. 5⅜ × 8½. (Available in U.S. only)　25032-6 Pa. $5.95

AUTHENTIC VICTORIAN DECORATION AND ORNAMENTATION IN FULL COLOR: 46 Plates from "Studies in Design," Christopher Dresser. Superb full-color lithographs reproduced from rare original portfolio of a major Victorian designer. 48pp. 9¼ × 12¼.　25083-0 Pa. $7.95

PRIMITIVE ART, Franz Boas. Remains the best text ever prepared on subject, thoroughly discussing Indian, African, Asian, Australian, and, especially, Northern American primitive art. Over 950 illustrations show ceramics, masks, totem poles, weapons, textiles, paintings, much more. 376pp. 5⅜ × 8.　20025-6 Pa. $6.95

SIDELIGHTS ON RELATIVITY, Albert Einstein. Unabridged republication of two lectures delivered by the great physicist in 1920–21. *Ether and Relativity* and *Geometry and Experience*. Elegant ideas in non-mathematical form, accessible to intelligent layman. vi + 56pp. 5⅜ × 8½.　24511-X Pa. $2.95

THE WIT AND HUMOR OF OSCAR WILDE, edited by Alvin Redman. More than 1,000 ripostes, paradoxes, wisecracks: Work is the curse of the drinking classes, I can resist everything except temptation, etc. 258pp. 5⅜ × 8½.　20602-5 Pa. $3.95

ADVENTURES WITH A MICROSCOPE, Richard Headstrom. 59 adventures with clothing fibers, protozoa, ferns and lichens, roots and leaves, much more. 142 illustrations. 232pp. 5⅜ × 8½.　23471-1 Pa. $3.95

PLANTS OF THE BIBLE, Harold N. Moldenke and Alma L. Moldenke. Standard reference to all 230 plants mentioned in Scriptures. Latin name, biblical reference, uses, modern identity, much more. Unsurpassed encyclopedic resource for scholars, botanists, nature lovers, students of Bible. Bibliography. Indexes. 123 black-and-white illustrations. 384pp. 6 × 9. 25069-5 Pa. $8.95

FAMOUS AMERICAN WOMEN: A Biographical Dictionary from Colonial Times to the Present, Robert McHenry, ed. From Pocahontas to Rosa Parks, 1,035 distinguished American women documented in separate biographical entries. Accurate, up-to-date data, numerous categories, spans 400 years. Indices. 493pp. 6½ × 9¼. 24523-3 Pa. $9.95

THE FABULOUS INTERIORS OF THE GREAT OCEAN LINERS IN HISTORIC PHOTOGRAPHS, William H. Miller, Jr. Some 200 superb photographs capture exquisite interiors of world's great "floating palaces"—1890's to 1980's: *Titanic, Ile de France, Queen Elizabeth, United States, Europa,* more. Approx. 200 black-and-white photographs. Captions. Text. Introduction. 160pp. 8⅞ × 11¾. 24756-2 Pa. $9.95

THE GREAT LUXURY LINERS, 1927–1954: A Photographic Record, William H. Miller, Jr. Nostalgic tribute to heyday of ocean liners. 186 photos of Ile de France, Normandie, Leviathan, Queen Elizabeth, United States, many others. Interior and exterior views. Introduction. Captions. 160pp. 9 × 12. 24056-8 Pa. $9.95

A NATURAL HISTORY OF THE DUCKS, John Charles Phillips. Great landmark of ornithology offers complete detailed coverage of nearly 200 species and subspecies of ducks: gadwall, sheldrake, merganser, pintail, many more. 74 full-color plates, 102 black-and-white. Bibliography. Total of 1,920pp. 8⅝ × 11¼. 25141-1, 25142-X Cloth. Two-vol. set $100.00

THE SEAWEED HANDBOOK: An Illustrated Guide to Seaweeds from North Carolina to Canada, Thomas F. Lee. Concise reference covers 78 species. Scientific and common names, habitat, distribution, more. Finding keys for easy identification. 224pp. 5⅜ × 8½. 25215-9 Pa. $5.95

THE TEN BOOKS OF ARCHITECTURE: The 1755 Leoni Edition, Leon Battista Alberti. Rare classic helped introduce the glories of ancient architecture to the Renaissance. 68 black-and-white plates. 336pp. 8⅜ × 11¼. 25239-6 Pa. $14.95

MISS MACKENZIE, Anthony Trollope. Minor masterpieces by Victorian master unmasks many truths about life in 19th-century England. First inexpensive edition in years. 392pp. 5⅜ × 8½. 25201-9 Pa. $7.95

THE RIME OF THE ANCIENT MARINER, Gustave Doré, Samuel Taylor Coleridge. Dramatic engravings considered by many to be his greatest work. The terrifying space of the open sea, the storms and whirlpools of an unknown ocean, the ice of Antarctica, more—all rendered in a powerful, chilling manner. Full text. 38 plates. 77pp. 9¼ × 12. 22305-1 Pa. $4.95

THE EXPEDITIONS OF ZEBULON MONTGOMERY PIKE, Zebulon Montgomery Pike. Fascinating first-hand accounts (1805–6) of exploration of Mississippi River, Indian wars, capture by Spanish dragoons, much more. 1,088pp. 5⅜ × 8½. 25254-X, 25255-8 Pa. Two-vol. set $23.90

A CONCISE HISTORY OF PHOTOGRAPHY: Third Revised Edition, Helmut Gernsheim. Best one-volume history—camera obscura, photochemistry, daguerreotypes, evolution of cameras, film, more. Also artistic aspects—landscape, portraits, fine art, etc. 281 black-and-white photographs. 26 in color. 176pp. 8⅜ × 11¼. 25128-4 Pa. $12.95

THE DORÉ BIBLE ILLUSTRATIONS, Gustave Doré. 241 detailed plates from the Bible: the Creation scenes, Adam and Eve, Flood, Babylon, battle sequences, life of Jesus, etc. Each plate is accompanied by the verses from the King James version of the Bible. 241pp. 9 × 12. 23004-X Pa. $8.95

HUGGER-MUGGER IN THE LOUVRE, Elliot Paul. Second Homer Evans mystery-comedy. Theft at the Louvre involves sleuth in hilarious, madcap caper. "A knockout."—Books. 336pp. 5⅜ × 8½. 25185-3 Pa. $5.95

FLATLAND, E. A. Abbott. Intriguing and enormously popular science-fiction classic explores the complexities of trying to survive as a two-dimensional being in a three-dimensional world. Amusingly illustrated by the author. 16 illustrations. 103pp. 5⅜ × 8½. 20001-9 Pa. $2.00

THE HISTORY OF THE LEWIS AND CLARK EXPEDITION, Meriwether Lewis and William Clark, edited by Elliott Coues. Classic edition of Lewis and Clark's day-by-day journals that later became the basis for U.S. claims to Oregon and the West. Accurate and invaluable geographical, botanical, biological, meteorological and anthropological material. Total of 1,508pp. 5⅜ × 8½.
21268-8, 21269-6, 21270-X Pa. Three-vol. set $25.50

LANGUAGE, TRUTH AND LOGIC, Alfred J. Ayer. Famous, clear introduction to Vienna, Cambridge schools of Logical Positivism. Role of philosophy, elimination of metaphysics, nature of analysis, etc. 160pp. 5⅜ × 8½. (Available in U.S. and Canada only) 20010-8 Pa. $2.95

MATHEMATICS FOR THE NONMATHEMATICIAN, Morris Kline. Detailed, college-level treatment of mathematics in cultural and historical context, with numerous exercises. For liberal arts students. Preface. Recommended Reading Lists. Tables. Index. Numerous black-and-white figures. xvi + 641pp. 5⅜ × 8½.
24823-2 Pa. $11.95

28 SCIENCE FICTION STORIES, H. G. Wells. Novels, *Star Begotten* and *Men Like Gods,* plus 26 short stories: "Empire of the Ants," "A Story of the Stone Age," "The Stolen Bacillus," "In the Abyss," etc. 915pp. 5⅜ × 8½. (Available in U.S. only)
20265-8 Cloth. $10.95

HANDBOOK OF PICTORIAL SYMBOLS, Rudolph Modley. 3,250 signs and symbols, many systems in full; official or heavy commercial use. Arranged by subject. Most in Pictorial Archive series. 143pp. 8⅛ × 11. 23357-X Pa. $5.95

INCIDENTS OF TRAVEL IN YUCATAN, John L. Stephens. Classic (1843) exploration of jungles of Yucatan, looking for evidences of Maya civilization. Travel adventures, Mexican and Indian culture, etc. Total of 669pp. 5⅜ × 8½.
20926-1, 20927-X Pa., Two-vol. set $9.90

DEGAS: An Intimate Portrait, Ambroise Vollard. Charming, anecdotal memoir by famous art dealer of one of the greatest 19th-century French painters. 14 black-and-white illustrations. Introduction by Harold L. Van Doren. 96pp. 5⅜ × 8½.
25131-4 Pa. $3.95

PERSONAL NARRATIVE OF A PILGRIMAGE TO ALMANDINAH AND MECCAH, Richard Burton. Great travel classic by remarkably colorful personality. Burton, disguised as a Moroccan, visited sacred shrines of Islam, narrowly escaping death. 47 illustrations. 959pp. 5⅜ × 8½.   21217-3, 21218-1 Pa., Two-vol. set $17.90

PHRASE AND WORD ORIGINS, A. H. Holt. Entertaining, reliable, modern study of more than 1,200 colorful words, phrases, origins and histories. Much unexpected information. 254pp. 5⅜ × 8½.   20758-7 Pa. $4.95

THE RED THUMB MARK, R. Austin Freeman. In this first Dr. Thorndyke case, the great scientific detective draws fascinating conclusions from the nature of a single fingerprint. Exciting story, authentic science. 320pp. 5⅜ × 8½. (Available in U.S. only)   25210-8 Pa. $5.95

AN EGYPTIAN HIEROGLYPHIC DICTIONARY, E. A. Wallis Budge. Monumental work containing about 25,000 words or terms that occur in texts ranging from 3000 B.C. to 600 A.D. Each entry consists of a transliteration of the word, the word in hieroglyphs, and the meaning in English. 1,314pp. 6⅞ × 10.
23615-3, 23616-1 Pa., Two-vol. set $27.90

THE COMPLEAT STRATEGYST: Being a Primer on the Theory of Games of Strategy, J. D. Williams. Highly entertaining classic describes, with many illustrated examples, how to select best strategies in conflict situations. Prefaces. Appendices. xvi + 268pp. 5⅜ × 8½.   25101-2 Pa. $5.95

THE ROAD TO OZ, L. Frank Baum. Dorothy meets the Shaggy Man, little Button-Bright and the Rainbow's beautiful daughter in this delightful trip to the magical Land of Oz. 272pp. 5⅜ × 8.   25208-6 Pa. $4.95

POINT AND LINE TO PLANE, Wassily Kandinsky. Seminal exposition of role of point, line, other elements in non-objective painting. Essential to understanding 20th-century art. 127 illustrations. 192pp. 6½ × 9¼.   23808-3 Pa. $4.50

LADY ANNA, Anthony Trollope. Moving chronicle of Countess Lovel's bitter struggle to win for herself and daughter Anna their rightful rank and fortune—perhaps at cost of sanity itself. 384pp. 5⅜ × 8½.   24669-8 Pa. $6.95

EGYPTIAN MAGIC, E. A. Wallis Budge. Sums up all that is known about magic in Ancient Egypt: the role of magic in controlling the gods, powerful amulets that warded off evil spirits, scarabs of immortality, use of wax images, formulas and spells, the secret name, much more. 253pp. 5⅜ × 8½.   22681-6 Pa. $4.00

THE DANCE OF SIVA, Ananda Coomaraswamy. Preeminent authority unfolds the vast metaphysic of India: the revelation of her art, conception of the universe, social organization, etc. 27 reproductions of art masterpieces. 192pp. 5⅜ × 8½.
24817-8 Pa. $5.95

CHRISTMAS CUSTOMS AND TRADITIONS, Clement A. Miles. Origin, evolution, significance of religious, secular practices. Caroling, gifts, yule logs, much more. Full, scholarly yet fascinating; non-sectarian. 400pp. 5⅜ × 8½.
23354-5 Pa. $6.50

THE HUMAN FIGURE IN MOTION, Eadweard Muybridge. More than 4,500 stopped-action photos, in action series, showing undraped men, women, children jumping, lying down, throwing, sitting, wrestling, carrying, etc. 390pp. 7⅞ × 10⅝.
20204-6 Cloth. $19.95

THE MAN WHO WAS THURSDAY, Gilbert Keith Chesterton. Witty, fast-paced novel about a club of anarchists in turn-of-the-century London. Brilliant social, religious, philosophical speculations. 128pp. 5⅜ × 8½. 25121-7 Pa. $3.95

A CEZANNE SKETCHBOOK: Figures, Portraits, Landscapes and Still Lifes, Paul Cezanne. Great artist experiments with tonal effects, light, mass, other qualities in over 100 drawings. A revealing view of developing master painter, precursor of Cubism. 102 black-and-white illustrations. 144pp. 8¾ × 6⅜. 24790-2 Pa. $5.95

AN ENCYCLOPEDIA OF BATTLES: Accounts of Over 1,560 Battles from 1479 B.C. to the Present, David Eggenberger. Presents essential details of every major battle in recorded history, from the first battle of Megiddo in 1479 B.C. to Grenada in 1984. List of Battle Maps. New Appendix covering the years 1967–1984. Index. 99 illustrations. 544pp. 6½ × 9¼. 24913-1 Pa. $14.95

AN ETYMOLOGICAL DICTIONARY OF MODERN ENGLISH, Ernest Weekley. Richest, fullest work, by foremost British lexicographer. Detailed word histories. Inexhaustible. Total of 856pp. 6½ × 9¼.
21873-2, 21874-0 Pa., Two-vol. set $17.00

WEBSTER'S AMERICAN MILITARY BIOGRAPHIES, edited by Robert McHenry. Over 1,000 figures who shaped 3 centuries of American military history. Detailed biographies of Nathan Hale, Douglas MacArthur, Mary Hallaren, others. Chronologies of engagements, more. Introduction. Addenda. 1,033 entries in alphabetical order. xi + 548pp. 6½ × 9¼. (Available in U.S. only)
24758-9 Pa. $11.95

LIFE IN ANCIENT EGYPT, Adolf Erman. Detailed older account, with much not in more recent books: domestic life, religion, magic, medicine, commerce, and whatever else needed for complete picture. Many illustrations. 597pp. 5⅜ × 8½.
22632-8 Pa. $8.50

HISTORIC COSTUME IN PICTURES, Braun & Schneider. Over 1,450 costumed figures shown, covering a wide variety of peoples: kings, emperors, nobles, priests, servants, soldiers, scholars, townsfolk, peasants, merchants, courtiers, cavaliers, and more. 256pp. 8⅜ × 11¼. 23150-X Pa. $7.95

THE NOTEBOOKS OF LEONARDO DA VINCI, edited by J. P. Richter. Extracts from manuscripts reveal great genius; on painting, sculpture, anatomy, sciences, geography, etc. Both Italian and English. 186 ms. pages reproduced, plus 500 additional drawings, including studies for *Last Supper, Sforza* monument, etc. 860pp. 7⅞ × 10¾. (Available in U.S. only) 22572-0, 22573-9 Pa., Two-vol. set $25.90

AMERICAN CLIPPER SHIPS: 1833–1858, Octavius T. Howe & Frederick C. Matthews. Fully-illustrated, encyclopedic review of 352 clipper ships from the period of America's greatest maritime supremacy. Introduction. 109 halftones. 5 black-and-white line illustrations. Index. Total of 928pp. 5⅜ × 8½.
25115-2, 25116-0 Pa., Two-vol. set $17.90

TOWARDS A NEW ARCHITECTURE, Le Corbusier. Pioneering manifesto by great architect, near legendary founder of "International School." Technical and aesthetic theories, views on industry, economics, relation of form to function, "mass-production spirit," much more. Profusely illustrated. Unabridged translation of 13th French edition. Introduction by Frederick Etchells. 320pp. 6⅛ × 9¼. (Available in U.S. only)
25023-7 Pa. $8.95

THE BOOK OF KELLS, edited by Blanche Cirker. Inexpensive collection of 32 full-color, full-page plates from the greatest illuminated manuscript of the Middle Ages, painstakingly reproduced from rare facsimile edition. Publisher's Note. Captions. 32pp. 9⅜ × 12¼.
24345-1 Pa. $4.50

BEST SCIENCE FICTION STORIES OF H. G. WELLS, H. G. Wells. Full novel *The Invisible Man*, plus 17 short stories: "The Crystal Egg," "Aepyornis Island," "The Strange Orchid," etc. 303pp. 5⅜ × 8½. (Available in U.S. only)
21531-8 Pa. $4.95

AMERICAN SAILING SHIPS: Their Plans and History, Charles G. Davis. Photos, construction details of schooners, frigates, clippers, other sailcraft of 18th to early 20th centuries—plus entertaining discourse on design, rigging, nautical lore, much more. 137 black-and-white illustrations. 240pp. 6⅛ × 9¼.
24658-2 Pa. $5.95

ENTERTAINING MATHEMATICAL PUZZLES, Martin Gardner. Selection of author's favorite conundrums involving arithmetic, money, speed, etc., with lively commentary. Complete solutions. 112pp. 5⅜ × 8½.
25211-6 Pa. $2.95

THE WILL TO BELIEVE, HUMAN IMMORTALITY, William James. Two books bound together. Effect of irrational on logical, and arguments for human immortality. 402pp. 5⅜ × 8½.
20291-7 Pa. $7.50

THE HAUNTED MONASTERY and THE CHINESE MAZE MURDERS, Robert Van Gulik. 2 full novels by Van Gulik continue adventures of Judge Dee and his companions. An evil Taoist monastery, seemingly supernatural events; overgrown topiary maze that hides strange crimes. Set in 7th-century China. 27 illustrations. 328pp. 5⅜ × 8½.
23502-5 Pa. $5.00

CELEBRATED CASES OF JUDGE DEE (DEE GOONG AN), translated by Robert Van Gulik. Authentic 18th-century Chinese detective novel; Dee and associates solve three interlocked cases. Led to Van Gulik's own stories with same characters. Extensive introduction. 9 illustrations. 237pp. 5⅜ × 8½.
23337-5 Pa. $4.95

4.20

This book is to be returned on or before
the last date stamped below.